AF581670

Transitioning to **Zero Hunger**

Transitioning to Sustainability Series: Volume 2

Series Editor: Manfred Max Bergman

Volumes in the series:

Volume 1: Transitioning to No Poverty
ISBN 978-3-03897-860-2 (Hbk);
ISBN 978-3-03897-861-9 (PDF)

Volume 2: Transitioning to Zero Hunger
ISBN 978-3-03897-862-6 (Hbk);
ISBN 978-3-03897-863-3 (PDF)

Volume 3: Transitioning to Good Health and Well-Being
ISBN 978-3-03897-864-0 (Hbk);
ISBN 978-3-03897-865-7 (PDF)

Volume 4: Transitioning to Quality Education
ISBN 978-3-03897-892-3 (Hbk);
ISBN 978-3-03897-893-0 (PDF)

Volume 5: Transitioning to Gender Equality
ISBN 978-3-03897-866-4 (Hbk);
ISBN 978-3-03897-867-1 (PDF)

Volume 6: Transitioning to Clean Water and Sanitation
ISBN 978-3-03897-774-2 (Hbk);
ISBN 978-3-03897-775-9 (PDF)

Volume 7: Transitioning to Affordable and Clean Energy
ISBN 978-3-03897-776-6 (Hbk);
ISBN 978-3-03897-777-3 (PDF)

Volume 8: Transitioning to Decent Work and Economic Growth
ISBN 978-3-03897-778-0 (Hbk);
ISBN 978-3-03897-779-7 (PDF)

Volume 9: Transitioning to Sustainable Industry, Innovation and Infrastructure
ISBN 978-3-03897-868-8 (Hbk);
ISBN 978-3-03897-869-5 (PDF)

Volume 10: Transitioning to Reduced Inequalities
ISBN 978-3-03921-160-9 (Hbk);
ISBN 978-3-03921-161-6 (PDF)

Volume 11: Transitioning to Sustainable Cities and Communities
ISBN 978-3-03897-870-1 (Hbk);
ISBN 978-3-03897-871-8 (PDF)

Volume 12: Transitioning to Responsible Consumption and Production
ISBN 978-3-03897-872-5 (Hbk);
ISBN 978-3-03897-873-2 (PDF)

Volume 13: Transitioning to Climate Action
ISBN 978-3-03897-874-9 (Hbk);
ISBN 978-3-03897-875-6 (PDF)

Volume 14: Transitioning to Sustainable Life below Water
ISBN 978-3-03897-876-3 (Hbk);
ISBN 978-3-03897-877-0 (PDF)

Volume 15: Transitioning to Sustainable Life on Land
ISBN 978-3-03897-878-7 (Hbk);
ISBN 978-3-03897-879-4 (PDF)

Volume 16: Transitioning to Peace, Justice and Strong Institutions
ISBN 978-3-03897-880-0 (Hbk);
ISBN 978-3-03897-881-7 (PDF)

Volume 17: Transitioning to Strong Partnerships for the Sustainable Development Goals
ISBN 978-3-03897-882-4 (Hbk);
ISBN 978-3-03897-883-1 (PDF)

Delwendé Innocent Kiba (Ed.)

Transitioning to **Zero Hunger**

Transitioning to Sustainability Series

MDPI • Basel • Beijing • Wuhan • Barcelona • Belgrade • Manchester • Tianjin • Tokyo • Cluj

EDITOR
Delwendé Innocent Kiba
Centre National de Recherche Scientifique
et Technologique (CNRST)
Ouagadougou, Burkina Faso

EDITORIAL OFFICE
MDPI
St. Alban-Anlage 66
4052 Basel, Switzerland

For citation purposes, cite each article independently as indicated below:

Author 1, and Author 2. 2023. Chapter Title. In *Transitioning to Zero Hunger*. Edited by Delwendé Innocent Kiba. Transitioning to Sustainability Series 2. Basel: MDPI, Page Range.

ISBN 978-3-03897-862-6 (Hbk)
ISBN 978-3-03897-863-3 (PDF)
ISSN: 2624-9324 (Print)
ISSN: 2624-9332 (Online)
doi:10.3390/books978-3-03897-863-3

Contents

	About the Editor	vii
	Contributors	ix
	Abstracts	xi
1	Preface to Transitioning to Zero Hunger DELWENDÉ INNOCENT KIBA	1
2	Challenges for Zero Hunger (SDG 2): Links with Other SDGs RAMONA LILE, MONICA OCNEAN AND IOANA MIHAELA BALAN	9
3	The Power of Social Capital to Address Structural Factors of Hunger GIAN L. NICOLAY	67
4	The Implications of Agroecology and Conventional Agriculture for Food Security and the Environment in Africa TERENCE EPULE EPULE AND ABDELGHANI CHEHBOUNI	95
5	Agroforestry: An Avenue for Resilient and Productive Farming through Integrated Crops and Livestock Production NUWANDHYA S. DISSANAYAKA, SHASHI S. UDUMANN, THARINDU D. NUWARAPAKSHA AND ANJANA J. ATAPATTU	115
6	Improving the Diversity of Native Edible Plants and Traditional Food and Agriculture Practices for Sustainable Food Security in the Future PERMANI C. WEERASEKARA AND ANGELIKA PLOEGER	137
7	Eliminating Hunger: Yam for Improved Income and Food Security in West Africa BEATRICE AIGHEWI, NORBERT MAROYA, ROBERT ASIEDU, DJANA MIGNOUNA, MORUFAT BALOGUN AND P. LAVA KUMAR	175
8	Coconut-Based Livestock Farming: A Sustainable Approach to Enhancing Food Security in Sri Lanka THARINDU D. NUWARAPAKSHA, SHASHI S. UDUMANN, NUWANDHYA S. DISSANAYAKA AND ANJANA J. ATAPATTU	197
9	Approaches to Limiting Food Loss and Food Waste IOANA MIHAELA BALAN, TEODOR IOAN TRASCA, IOAN BRAD, NASTASIA BELC, CAMELIA TULCAN, BOGDAN PETRU RADOI AND ALEXANDRU ERNE RINOVETZ	215

About the Editor

Delwendé Innocent Kiba, former Minister of Agriculture and Animal Resources, is a soil scientist at the Institut de l'Environnement et de Recherches Agricoles (INERA) in Burkina Faso. He completed his PhD on Rural Development at the Polytechnic University of Bobo-Dioulasso (UPB) in 2012. Thereafter, he completed a postdoc in the Group of Plant Nutrition at ETH Zurich in Switzerland. From 2014 to 2021, he was the overall coordinator of a project on sustainable soil use in yam systems in West Africa (YAMSYS project). Kiba has co-authored numerous publications on integrated soil fertility management in Africa and has given numerous communications around the world. He has trained numerous students (BSc, Ing., MSc, PhD). In addition to his work as a researcher, he was the general secretary of the Soil Science Society of Burkina Faso from 2014 to 2017 and the president of this society since 2017. Kiba's research has been distinguished several times, including the scientific prize of the Belgian Development Cooperation in 2007, the Swiss Government Excellence Scholarship in 2007, scientific prizes at the Forum of Science Innovation and Technology (FRSIT) in Burkina Faso, finalist at the Forum for Agricultural Research in Africa (FARA) scientific competition in 2010, and a co-recipient of the CSRS-Fond Eremitage award in 2019 for his active participation in partnership research. He is affiliated with the African Academy of Science (2021–2025).

Contributors

ABDELGHANI CHEHBOUNI
PhD, International Water Research Institute (IWRI), Mohammed VI Polytechnic University (UM6P), Benguerir, Morocco.

ALEXANDRU ERNE RINOVETZ
PhD, Faculty of Food Engineering, University of Life Sciences "King Michael I" from Timisoara, Romania.

ANGELIKA PLOEGER
Prof., Specialized Partnerships in Sustainable Food Systems and Food Sovereignty, Faculty of Organic Agricultural Sciences, University of Kassel, Witzenhausen, Germany.

ANJANA J. ATAPATTU
Dr., Agronomy Division, Coconut Research Institute of Sri Lanka, Lunuwila, Sri Lanka.

BEATRICE AIGHEWI
Assoc. Prof., Department of Tropical Agriculture, International Institute of Tropical Agriculture, Ibadan, Nigeria.

BOGDAN PETRU RADOI
PhD, Faculty of Food Engineering, University of Life Sciences "King Michael I" from Timisoara, Romania.

CAMELIA TULCAN
Prof., Faculty of Engineering and Applied Technologies, University of Life Sciences "King Michael I" from Timisoara, Romania.

DJANA MIGNOUNA
PhD, International Institute of Tropical Agriculture (IITA), Cotonou, Republic of Benin.

NUWANDHYA S. DISSANAYAKA
MSc (Reading), Agronomy Division, Coconut Research Institute of Sri Lanka, Lunuwila, Sri Lanka.

GIAN L. NICOLAY
MSc, Africa coordinator, Research Institute of Organic Agriculture (FiBL), Frick, Switzerland.

IOANA MIHAELA BALAN
Assoc. Prof., Faculty of Management and Rural Tourism, University of Life Sciences "King Michael I" from Timisoara, Timisoara, Romania.

IOAN BRAD
Prof., Faculty of Management and Rural Tourism, University of Life Sciences "King Michael I" from Timisoara, Romania.

MONICA OCNEAN
Dr., University of Life Sciences "King Michael I" from Timisoara, Romania.

MORUFAT BALOGUN
Assoc. Prof., University of Ibadan, Ibadan, Nigeria;
International Institute of Tropical Agriculture (IITA), Ibadan, Nigeria.

NASTASIA BELC
Assoc. Prof., Faculty of Biotechnology, University of Agronomy and Veterinary Medicine, Bucharest, Romania.

NORBERT MAROYA
Dr., International Institute of Tropical Agriculture (IITA), Cotonou, Republic of Benin.

PERMANI C. WEERASEKARA
Dr., Specialized Partnerships in Sustainable Food Systems and Food Sovereignty, Faculty of Organic Agricultural Sciences, University of Kassel, Witzenhausen, Germany.

P. LAVA KUMAR
PhD, International Institute of Tropical Agriculture (IITA), Ibadan, Nigeria.

RAMONA LILE
Prof., Faculty of Economics, "Aurel Vlaicu" University of Arad, Arad, Romania.

ROBERT ASIEDU
Dr., International Institute of Tropical Agriculture (IITA), Ibadan, Nigeria.

SHASHI S. UDUMANN
MSc (Reading), Agronomy Division, Coconut Research Institute of Sri Lanka, Lunuwila, Sri Lanka.

THARINDU D. NUWARAPAKSHA
M.AETM, Agronomy Division, Coconut Research Institute of Sri Lanka, Lunuwila, Sri Lanka.

TEODOR IOAN TRASCA
Prof., Faculty of Food Engineering, University of Life Sciences "King Michael I" from Timisoara, Romania.

TERENCE EPULE EPULE
Dr., International Water Research Institute (IWRI), Mohammed VI Polytechnic University (UM6P), Benguerir, Morocco.

Abstracts

Preface to Transitioning to Zero Hunger
by Delwendé Innocent Kiba

Food insecurity remains a global concern, with challenges exacerbated by social and health crises, namely the COVID-19 pandemic and geopolitical conflicts. Experts agree that food security requires coordinated efforts for a transition to zero hunger. Despite technological advances, agrifood systems still face significant hidden costs linked to environmental, social, and health problems. Efforts to address food security are aligned with the Sustainable Development Goals (SDGs) adopted in 2015, but the FAO estimates that nearly 670 million people will still suffer from hunger in 2030. This book explores the quest for a world with access to sufficient, safe, and nutritious food, focusing on the need for inclusive, resilient, and sustainable agrifood systems. It analyzes the drivers of hunger and the sustainability of food production, offering pathways for a transition to zero hunger. The chapters examine the links between SDG 2 and the other SDGs, the role of social capital, and agroecology, presenting case studies on agroforestry, agrobiodiversity, yam farming in West Africa, livestock farming integrated with coconut production in Sri Lanka, and approaches to limit food waste.

Challenges for Zero Hunger (SDG 2): Links with Other SDGs
by Ramona Lile, Monica Ocnean, and Ioana Mihaela Balan

The Sustainable Development Goals (SDGs), ratified by the United Nations General Assembly in September 2015, embody a set of 17 objectives designed to address the world's most urgent challenges. SDG 2: Zero Hunger is intricately linked with all other SDGs, aiming not only to eliminate global hunger but also to collectively achieve the broader spectrum of all other SDGs. In this context, it becomes imperative to address and resolve issues encompassing poverty, health, education, inequality, and climate change in a comprehensive manner, while fostering equitable development. SDG 2: Zero Hunger centers on the elimination of hunger and malnutrition, promoting sustainable agricultural practices, and enhancing productivity and incomes for smallholder farmers and food producers, particularly in developing nations. Accomplishing these targets hinges on the development of resilient food systems, the promotion of innovative agricultural technologies, including those geared towards climate resilience. SDG 2: Zero Hunger encounters multifaceted challenges arising from climate change impacts, rapid urbanization, and the imperative to foster sustainable agricultural practices while reducing disparities. These challenges are inherently intertwined with the goals of the broader SDGs framework, necessitating cooperative efforts, innovative approaches, and concerted actions to guarantee global food security and establish a sustainable and equitable future for all of humanity. This book chapter underscores the undeniable direct and indirect links of SDG 2: Zero Hunger with the entire spectrum of SDGs, substantiated through tangible examples, but also the deficiencies and slow pace of progress. It also underscores the significance of a holistic and integrated approach, emphasizing the need to address SDG 2 and its complementary objectives in a synergistic manner, thereby facilitating tangible and lasting progress towards a more promising future.

The Power of Social Capital to Address Structural Factors of Hunger

by Gian L. Nicolay

This essay contributes to the theory of the current crises of the world food system and agriculture, including persistent hunger. It is organized into seven chapters and develops the critical importance of social capital in ending hunger. The introduction highlights the importance of a theoretical understanding of this issue to address the well-known symptoms under the guidance of the FAO. Then, the commonly agreed upon five groups of structural factors of hunger are recalled: poverty; wars and pandemics; gender, age, and race; divided societies; and finally capitalist-driven economies including land grabbing. Thirdly, the concept of social capital is proposed as related to social networks and social systems, and the consequences of its neglect as a hunger parameter are explained. Agroecology, often considered the solution since 2008, is critically analyzed and compared with the food regime based on industrial agriculture. These two regimes are confronted with a third method, applying the morphological analyses invented by Zwicky. The surprising results are further developed into proposals on how social capital can be created and used to end hunger. The essay develops around the main discourses since the IAASTD report, the food crises of 2008, and the required transformation into more sustainable forms. Social science and the concepts of social systems are essential in this narrative. We see the underdeveloped social capital, particularly social networks and other local institutions related to national policies at the local and rural levels, as a critical parameter and indicator to predict hunger or food and nutrition insecurity. Empirical studies and experiments from the author's research and work in Africa support this short and dense essay, hoping to contribute to a better understanding of ending hunger before 2050.

The Implications of Agroecology and Conventional Agriculture for Food Security and the Environment in Africa

by Terence Epule Epule and Abdelghani Chehbouni

As global climate continues to change, changes need to be made in our production systems to ensure global food production. These constraints are daunting in Africa, as Africa is the most vulnerable region to climate change and variability. Agroecology provides a unique opportunity for Africa to achieve the twin challenges of food security and environmental resilience. This chapter aims at examining the relative contributions of agroecology and conventional agriculture towards resilient food security in Africa. The chapter examines the theoretical foundations and components of these two paradigms as well as their contributions to food security in Africa. This chapter also examines the likely benefits and challenges associated with these systems and discusses in an integrated manner which of these options offers the most likely resilient agricultural revolution for Africa. The methodology is based on a bibliometric review of publications in the grey and peer-reviewed literature on this subject. The compendium of 49 suitable studies was culled through search engines such as Google Scholar, Scopus, SCI, and ISI Web of Science. It is observed that agroecology needs more valorization to be able to match the yields of conventional agriculture in Africa. Since agriculture in Africa is mostly in the hands of smallholders, production is generally under natural conditions driven by limited access to conventional production inputs. Agroecology will require inputs from conventional production to be able to sustain production, except the system is valorized.

Agroforestry: An Avenue for Resilient and Productive Farming through Integrated Crops and Livestock Production

by Nuwandhya S. Dissanayaka, Shashi S. Udumann, Tharindu D. Nuwarapaksha and Anjana J. Atapattu

The global population is expected to reach 9.7 billion by 2050. With more mouths to feed, achieving zero hunger becomes even more crucial to ensure that everyone has access to sufficient, safe, and nutritious food. Resilient and productive farming practices are crucial in attaining this objective and aligning with Sustainable Development Goal 2. By enhancing food security, promoting sustainability, supporting rural development, adapting to climate change, and conserving biodiversity, resilient and productive farming becomes an essential pathway towards zero hunger, benefiting both present and future generations. Agroforestry is a versatile, resilient farming approach that can be easily adapted to most of the cropping systems worldwide. It involves combining trees, crops, and/or livestock more efficiently and effectively to maximize land utilization and food production with lower environmental impacts and financial costs. This chapter provides an overview of different agroforestry systems suitable for resilient farming. It explores criteria for selecting and managing appropriate plant and animal species in agroforestry systems, highlighting the potential benefits in terms of land utilization and food production while minimizing financial costs and environmental impact. The chapter also explores the role of agroforestry in biodiversity preservation, climate change adaptation and mitigation, economic feasibility, and scaling up for resilient and productive farming.

Improving the Diversity of Native Edible Plants and Traditional Food and Agriculture Practices for Sustainable Food Security in the Future

by Permani C. Weerasekara and Angelika Ploeger

By the year 2030, agriculture will have to provide the food and nutrition requirements of some eight billion people. These include eradicating hunger, improving access to food, ending all forms of malnutrition, promoting sustainable agriculture, and preserving food diversity. Food security is one of the global challenges. Simultaneously, research and development are focused on improving the productivity of a small number of existing crops that will improve global food production instead of increasing the diversity of crops. The result is the loss of agrobiodiversity. Humans cultivate about 150 of the estimated 30,000 species of edible plants worldwide, and most of our diets consist of just 30 species. New commercial crops and local wild plants can diversify global food production and better allow local acclimation to the diverse environment humans inhabit. We consider the values of and advantages and barriers to using local traditional food plants and knowledge in Sri Lanka. Also, we examine the missed opportunity to commercially produce local wild plants in Sri Lanka. We examine how wild species have been determined to improve crop varieties and where efforts must be concentrated to harness their value in the future. This chapter aims to improve the use of traditional food plants through the diversification of food for sustainable food production and practices in Sri Lanka. This will benefit fighting hunger and improve sustainable biodiversity.

Eliminating Hunger: Yam for Improved Income and Food Security in West Africa

by Beatrice Aighewi, Norbert Maroya, Robert Asiedu, Djana Mignouna, Morufat Balogun and P. Lava Kumar

Yam, *Dioscorea* spp., is a valuable vegetatively propagated crop grown in many parts of the tropics. In West Africa, the species *Dioscorea rotundata* is a nutritious staple and provides food security and a means of livelihood to millions of people. Yam is produced mainly by smallholder farmers using local landraces with limited inputs. Increased annual production is attained by increasing the area while productivity is low and stagnated. Significant contributors to the low productivity include unavailability, high cost, poor quality of planting material, nematode and viral infections, and declining soil fertility. The multiplication ratio of yam in traditional production methods is low (1:3). Seed to replant the same size of field harvested consumes about a third of the total production, i.e., about 23.6 million tonnes out of 70.8 million tonnes of the annual production of the West African sub-region are reserved for planting the next crop. Improving the seed yam multiplication ratio and productivity will improve the availability of more yams for food. The initiative "Yam Improvement for Income and Food Security in West Africa (YIIFSWA)" has developed new strategies for improved propagation of quality yam planting materials and increased the multiplication ratio to 1:300 using nodal vine cuttings from plants produced in hydroponic systems instead of tubers, thereby releasing more tubers for food use. By using improved yam varieties with good agronomic practices as well as nematode and viral disease management, the productivity of yam is improved. These improvements have great potential to enhance food security and alleviate hunger and poverty.

Coconut-Based Livestock Farming: A Sustainable Approach to Enhancing Food Security in Sri Lanka

by Tharindu D. Nuwarapaksha, Shashi S. Udumann, Nuwandhya S. Dissanayaka and Anjana J. Atapattu

Coconut (*Cocos nucifera* L.) cultivation in Sri Lanka holds a significant position in the agricultural landscape of the country, contributing to the social, economic, and cultural lives of millions. Coconut-based livestock farming is an innovative and sustainable approach that combines coconut cultivation with livestock rearing to enhance food security and promote environmental sustainability. This chapter provides an overview of the potential benefits, challenges, and relevance of coconut-based livestock farming systems in achieving the sustainable development goals (SDGs), particularly in relation to food security in Sri Lanka. Coconut–livestock integration is a farming system that aims to establish a synergistic relationship between livestock and coconut by integrating them on the same land. This approach can be implemented in various forms, including agroforestry, silvopastoral systems, mixed crop–livestock systems, and integrated crop–livestock systems. The cultivation of forage crops in coconut plantations has been identified as a substantial agricultural endeavor, presenting opportunities for sustainable practices and livestock development. Fodder production and conservation on coconut land, particularly through silage, are explored as crucial components of coconut-based livestock farming. This chapter underscores the role of this integrated approach in achieving the SDGs related to poverty eradication, sustainable agriculture, responsible consumption, climate change mitigation, and biodiversity conservation. Moreover, these systems play a key role in achieving SDG 2 (Zero Hunger) and its associated objectives, while also contributing to SDG 8, SDG 12, SDG 13, SDG 15, and SDG 17. The study concludes by highlighting the potential of coconut-based livestock farming to contribute to the SDGs and promote sustainable and inclusive agricultural practices in Sri Lanka.

Approaches to Limiting Food Loss and Food Waste

by Ioana Mihaela Balan, Teodor Ioan Trasca, Ioan Brad, Nastasia Belc, Camelia Tulcan, Bogdan Petru Radoi and Alexandru Erne Rinovetz

Considering the general context of food loss and waste, the year 2011 marked a significant turning point, with data published by the Food and Agriculture Organization of the United Nations (FAO) estimating that more than a third of the world's food production is lost or wasted throughout the food chain. Alarmingly, this situation persisted for over a decade, until 2022, showing a regressive trajectory rather than improvement. This trend has had negative consequences, impacting economic, social, and environmental conditions, although an exact quantification of its effects remains elusive at present. Within this framework, the tandem challenges of food loss and food waste have emerged as important issues within global food systems, perpetuating a cycle of generating substantial volumes of edible food waste annually. The current chapter introduces a holistic approach designed to address the intricate facets of food loss and food waste across all stages of the food chain. In this context, this chapter proposes key and complementary measures aimed at mitigating these negative effects within relevant stages of the food chain. While the chapter does not propose to offer an exhaustive analysis, it nonetheless synthesizes the worldwide scenario, supplemented by a detailed illustration of the situation in Romania as a representative model. The research methodology involved both the examination of external data and the authors' own published data. The chapter's overarching conclusion underscores the resounding significance: in the context of Sustainable Development Goal 2—Zero Hunger, the reduction in food loss and food waste emerges as a solution for increasing quantities of available food for global population. This approach holds a dual boon, benefiting the environment by reducing water and land resource consumption and subsequently reducing greenhouse gas emissions. The outcomes will provide increased productivity, catalyze economic growth, and produce more sustainable societies.

Preface to Transitioning to Zero Hunger

Delwendé Innocent Kiba

1. Introduction

Food insecurity is a worldwide concern. Today, the quest for a world where everyone has access to sufficient, safe, and nutritious food has never been so crucial. In 2021, around 2.3 billion people were food-insecure, seriously affecting around 11.7% of the world's population (FAO et al. 2022). Food insecurity has been exacerbated by the social and health crises the world has been experiencing since 2019. The COVID-19 crisis and the Russia–Ukraine conflict have fundamentally affected food supply both for agricultural products and for the inputs used to produce them. These crises have disrupted supply chains, limited access to markets, and increased the vulnerability of marginalized populations (Ben Hassen and El Bilali 2022).

Experts agree that food security can only be achieved through inclusive, resilient, and sustainable agrifood systems (FAO 2023). Agrifood systems go beyond simple access to food and encompass the entire food chain, including the production, processing, distribution, and consumption of agricultural and food products (Neik et al. 2023). Agrifood systems are weakened by climate change, the loss of biodiversity, and rising poverty. Although advancements in technology offer promising avenues to enhance food production while minimizing environmental impact (Foley et al. 2011), agrifood systems still generate significant hidden costs linked to environmental, social, and health problems, as stated in the 2023 report of the FAO.

Since the adoption of the SDGs in 2015, efforts and initiatives to promote sustainable agriculture, ensure food and nutrition security and eradicate hunger have been deployed around the world. Today, there are many signs that the "zero hunger" objective set out in the Sustainable Development Goals (SDGs) will not be achieved. The FAO estimates that nearly 670 million people will still be suffering from hunger in 2030. This was made clear at the UN Summit on Food Systems in 2021, where several nations made a firm commitment to implement policies aimed at transforming agrifood systems.

This book analyzes the factors behind hunger and the sustainability of food production, and highlights approaches aimed at a transition to zero hunger.

Chapter 2 deals exhaustively with the links between SDG 2 and the other SDGs, in particular the global resolution of problems such as poverty, health, education, and climate change, while Chapter 3 highlights the essential role of social capital in combating the structural factors of hunger and provides a critical analysis of agroecology. The analyses of agroecology and conventional food production are expanded in Chapter 4. Chapters 5 and 6 deal with the resilience and productivity of production systems and their sustainability, with a particular focus on agroforestry and agrobiodiversity. Chapters 7 and 8 deal with case studies of yam farming in West Africa and coconut-based livestock farming in Sri Lanka, respectively, as contributions to achieving food security and improving people's incomes. Finally, in Chapter 9, approaches aimed at limiting food waste are presented as solutions for achieving the "Zero Hunger" objective while contributing to environmental sustainability.

2. Highlights

2.1. Pathways to Zero Hunger

SDG 2 aims to eradicate hunger and ensure food and nutritional security by 2030. Achieving this goal necessarily involves a phase of promoting sustainable food production. This transition phase towards zero hunger requires tackling various issues relating to the production, distribution, access, and use of foodstuffs. The transition to zero hunger will undoubtedly require coordinated efforts at the national, regional, and international level, with complex and multi-faceted challenges to reconcile increased agricultural productivity with environmental protection and global social and economic objectives (Okello et al. 2021). To this must be added efforts to guarantee physical and economic access to sufficient, healthy, and nutritious food, considering individual food preferences. Today, more than ever, it is necessary to integrate nutrition into food production, as malnutrition is recognized as a global phenomenon affecting both underdeveloped and developed countries (Beyerlee and Fanzo 2019). It is important to consider indigenous food systems, particularly the development of traditional crops that are often best adapted to their biophysical, socio-economic, and cultural contexts, but which have unfortunately been neglected until now (FAO 2022). In addition, the production of healthy, balanced, and sufficient food requires access to innovations and technologies such as precision agriculture, digital technologies, biotechnologies, and innovative solutions for the storage and transport of foodstuffs. Access to innovations does not guarantee their adoption by stakeholders. The adoption of innovations requires a good connection between research and development through a transdisciplinary approach that enables them to be co-created and correspond to the real needs of stakeholders. (Kiba et al. 2020; Jacobi et al. 2022). Participatory and adaptive solutions that consider local institutional capacities, the diversification of agroecosystems and ecological management, and the quality of local diets should therefore be favored (Blesh et al. 2019). Sustainable land management is essential if the world is to achieve food security. Soil, an essential support for agricultural production, is being degraded at an exponential rate, although the extent and impact of this degradation cannot be accurately assessed. It should be possible to develop a global approach to disentangle the natural and anthropogenic causes of soil degradation, based on ecological approaches to production and remote sensing (Bindraban et al. 2012).

None of the approaches toward zero hunger are possible without eradicating poverty (SDG 1) and inequality (SDG 10) and improving the livelihood of people (SDG 3). Finally, policy and governance around trade, land tenure, soil health, and regulations that affect food production and distribution are necessary, as are good partnerships between governments, non-governmental organizations, businesses, and the research community to share the necessary knowledge and resources.

2.2. Case Studies

In Chapter 2, Lile et al. explore the complex relationship between SDG 2 and the other SDGs to identify direct and indirect links. They show that SDG 2 does not stand alone but is instead closely linked to all the other SDGs. To achieve zero hunger, the authors stress the urgent need to find answers to the issues of poverty, health, education, inequality, and climate change in a comprehensive manner. They point out that given the interconnections between the different SDGs, zero hunger cannot be achieved through isolated efforts, but rather through a synergy of actions between government agencies, international organizations, civil society, private companies, and other stakeholders, each naturally playing its role according to its area of expertise. The authors suggest that policies and programs to tackle the root causes of hunger should be formulated transparently and integrated into national development programs that can be regularly monitored and evaluated.

In Chapter 3, Gian L. Nicolay examines the essential role of social capital in combating the structural factors contributing to hunger. The author analyzes the drivers of hunger, including poverty, wars, pandemics, climate change, gender, age, race, societal divisions, and capitalism. He shows that social capital is a key parameter for predicting and solving the problem of hunger. These include extension systems, agencies that link farmers to markets and external agencies, innovation platforms, farmer field schools, cooperatives, and business groups. According to the author, these social organizations enable sustainable intensification to succeed as an important element of the food system in low-income countries and improve economic performance. The author also makes a critical analysis of agroecology, comparing it with industrial agriculture, and proposes a morphological analysis and solutions for a sustainable food production.

In Chapter 4, Epule and Chehbouni discuss the implications of agroecology and conventional agriculture for food security in Africa. They highlight the fact that the continent faces the daunting challenges of climate change and variability and examine the contributions, benefits, and challenges associated with agroecology and conventional agriculture. The authors stress the need to promote agroecology so that it achieves the same yields as conventional agriculture. However, they recognize the difficulty of getting out of the agroecology/conventional agriculture dilemma, given the involvement of several factors in the African context. Looking ahead, the authors suggest innovations to make conventional agriculture cleaner and more sustainable, the multiplication of pilot studies on agroecology, and a political approach that is both bottom-up and top-down to guarantee the success of the various initiatives.

The study by Dissanayaka et al. (Chapter 5) presents agroforestry—combining trees, crops, and animals—as an important multi-purpose approach for the productivity and resilience of production systems. They highlight different agroforestry practices and the potential benefits in terms of land use, food production, biodiversity conservation, and adaptation to climate change. Factors such as site selection and planning, component selection, planting, system management and harvesting, and the use of products are presented as key elements in the implementation of agroforestry, although these elements may vary according to the context and objectives. Finally, the authors discuss the limitations of agroforestry systems and suggest ways of improving them.

Achieving food and nutritional security requires crop diversification, as demonstrated by the authors Weerasekara and Plooger from Sri Lanka (Chapter 6). In this chapter, the importance of agrobiodiversity and the need to diversify global food production are discussed, as are the risks of losing this agrobiodiversity. The authors recommend the use of local traditional food plants to achieve food and nutritional security. The study identified 85 species of food plants of great importance. Traditional food preparation and preservation methods are presented and discussed. From the authors' analysis, it emerges that traditional local crops are less costly and more environmentally friendly sources of food, since they are well adapted to unfavorable climatic and edaphic conditions and have good resistance to pests and diseases.

In Chapter 7, Aighewi et al. show the potential of a traditional crop such as yam for achieving food security and improving household incomes in West

Africa. They also outline the challenges faced by yam farmers, including poor seed quality. The authors present an initiative to improve yam production by setting up formal quality seed production systems as part of the project 'Yam Improvement for Income and Food Security in West Africa (YIIFSWA)'. This initiative has led to the production of certified yam seed and the development of yam seed markets. Virus detection tools and technologies for eliminating infected sources, as well as high-ratio propagation technologies, are presented as a strategy for improving yam seed quality and productivity, thereby contributing to food security and reducing household poverty.

In Chapter 8, Nuwarapaksha et al. show the importance of integrating coconut cultivation and livestock production in the Sri Lankan context as an innovative and sustainable approach offering mutual benefits, including increased productivity, improved soil health, and reduced dependence on mineral fertilizers. The authors argue that although this system provides livelihoods for farmers while conserving natural resources, further research is needed to identify the most appropriate animal and plant species.

Achieving zero hunger requires good management to minimize food waste. Chapter 9 by Balan et al. looks at limiting food loss and waste, a pervasive problem that persists despite global efforts. The authors map food losses around the world. They highlight the importance of reducing food loss and waste in the context of SDG 2 and propose key measures to mitigate these challenges at all stages of the food chain, including production, storage, processing, distribution, retail and food services, and household losses. Furthermore, the authors believe that it is essential to identify and assess the factors contributing to food waste to effectively tackle the problem. This chapter provides useful guidelines for authorities, businesses, organizations, and consumers to help them adopt sustainable practices and promote effective food management.

3. Conclusions

The transition to zero hunger calls for innovation throughout the chain of the agrifood system. This book has examined the links between SDG 2 and the other SDGs, as well as the role of social capital, agroecology, agroforestry, crop diversification, innovative farming systems, and reducing food losses as essential elements in achieving food and nutrition security. The book provides analyses and solutions from a variety of authors that can be used by policy makers, researchers, and practitioners engaged in the global effort to combat food and nutrition insecurity. The discussions highlight the need for holistic and collaborative approaches to address the complex web of challenges associated with achieving SDG 2 and related goals.

Funding: This research received no external funding.

Conflicts of Interest: The author declares no conflict of interest.

References

Ben Hassen, Tarek, and Hamid El Bilali. 2022. Impacts of the Russia-Ukraine War on Global Food Security: Towards More Sustainable and Resilient Food Systems? *Foods* 11: 2301. [CrossRef] [PubMed]

Beyerlee, Derek, and Jessica Fanzo. 2019. The SDG of zero hunger 75 years on: Turning full circle on agriculture and nutrition. *Global Food Security* 21: 52–59. [CrossRef]

Bindraban, Prem S., Marijn Van Der Velde, Liming Ye, Maurits van den Berg, Simeon Materechera, Delwendé Innocent Kiba, Lulseged Tamene, Kristín Vala Ragnarsdóttir, Raymond Jongschaap, Marianne Hoogmoed, and et al. 2012. Assessing the impact of soil degradation on food production. *Current Opinion in Environmental Sustainability* 4: 478–88. [CrossRef]

Blesh, Jennifer, Lesli Hoey, Andrew D. Jones, Harriet Friedmann, and Ivette Perfecto. 2019. Development pathways toward "zero hunger". *World Development* 118: 1–14. [CrossRef]

FAO. 2022. The Future of Food and Agriculture: Drivers and Triggers for Transformation. Available online: https://doi.org/10.4060/cc0959en (accessed on 22 October 2023).

FAO. 2023. In Brief to The State of Food and Agriculture 2023: Revealing the True Cost of Food to Transform Agrifood Systems. Available online: https://doi.org/10.4060/cc7937en (accessed on 24 October 2023).

FAO, IFAD, UNICEF, WFP, and WHO. 2022. In Brief to The State of Food Security and Nutrition in the World. Repurposing Food and Agricultural Policies to Make Healthy Diets More Affordable. Available online: https://doi.org/10.4060/cc0640en (accessed on 22 October 2023).

Foley, Jonathan A., Navin Ramankutty, Kate A. Brauman, Emily S. Cassidy, James S. Gerber, Matt Johnston, Nathaniel D. Mueller, Christine O'Connell, Deepak K. Ray, Paul C. West, and et al. 2011. Solutions for a cultivated planet. *Nature* 478: 337–42. [CrossRef]

Jacobi, J., A. Llanque, S. M. Mukhovi, E. Birachi, P. von Groote, R. Eschen, I. Hilber-Schöb, D. I. Kiba, E. Frossard, and C. Robledo-Abad. 2022. Transdisciplinary co-creation increases the utilization of knowledge from sustainable development research. *Environmental Science & Policy* 129: 107–15. [CrossRef]

Kiba, Delwendé Innocent, Valérie Kouamé Hgaza, Beatrice Aighewi, Sévérin Aké, Dominique Barjolle, Thomas Bernet, Lucien N. Diby, Léa Jeanne Ilboudo, Gian Nicolay, Esther Oka, and et al. 2020. A Transdisciplinary Approach for the Development of Sustainable Yam (*Dioscorea* sp.) Production in West Africa. *Sustainability* 12: 4016. [CrossRef]

Neik, Ting Xiang, Kadambot H. M. Siddique, Sean Mayes, David Edwards, Jacqueline Batley, Tafadzwanashe Mabhaudhi, Beng Kah Song, and Festo Massawe. 2023. Diversifying agrifood systems to ensure global food security following the Russia–Ukraine crisis. *Frontiers in Sustainable Food Systems* 7: 1124640. [CrossRef]

Okello, Moses, Jimmy Lamo, Mildred Ochwo-Ssemakula, and Francis Onyilo. 2021. Challenges and innovations in achieving zero hunger and environmental sustainability through the lens of sub-Saharan Africa. *Outlook on Agriculture* 50: 141–47. [CrossRef]

Challenges for Zero Hunger (SDG 2): Links with Other SDGs

Ramona Lile, Monica Ocnean and Ioana Mihaela Balan

1. Introduction: SDG 2 and Its Integral Connection to the Sustainable Development Goals Framework

Adopted by the United Nations General Assembly in September 2015, the Sustainable Development Goals (SDGs) comprise a comprehensive framework of 17 interconnected objectives (SDG 1: No Poverty, SDG 2: Zero Hunger, SDG 3: Good Health and Well-being, SDG 4: Quality Education, SDG 5: Gender Equality, SDG 6: Clean Water and Sanitation, SDG 7: Affordable and Clean Energy, SDG 8: Decent Work and Economic Growth, SDG 9: Industry, Innovation and Infrastructure, SDG 10: Reduced Inequality, SDG 11: Sustainable Cities and Communities, SDG 12: Responsible Consumption and Production, SDG 13: Climate Action, SDG 14: Life Below Water, SDG 15: Life on Land, SDG 16: Peace, Justice and Strong Institutions, SDG 17: Partnerships to achieve the Goals). These goals collectively address some of the most important global challenges, spanning from poverty and hunger to inequality, climate change, and sustainable economic growth. The inception of the SDGs can be traced back to the 2012 United Nations Summit on Sustainable Development, commonly referred to as Rio+20, which underscored the need for a unified approach to tackle these pressing issues (UN The 17 Goals n.d.; Otto-Zimmermann 2012).

The process of crafting the SDGs was characterized by inclusivity, involving governments, the private sector, civil society, and diverse stakeholders on a global scale. This collaborative endeavor encompassed extensive consultations, negotiations, and rigorous studies to pinpoint critical concerns and delineate specific objectives to combat them. The culmination of this process was the unanimous adoption of the 17 Sustainable Development Goals in September 2015 by the United Nations General Assembly, serving as a clarion call for collective action to achieve these aims by 2030 (UN Transforming Our World: The 2030 Agenda for Sustainable Development n.d.).

One pivotal goal within this framework is SDG 2: Zero Hunger, which is intricately intertwined with all the other SDGs, forming a nexus of interdependence that underscores the significance of these connections. SDG 2 is dedicated to eliminating hunger and malnutrition, fostering sustainable agriculture, and uplifting the livelihoods of smallholder farmers, particularly in developing nations. The attainment of SDG 2 necessitates a holistic approach that addresses an array of challenges, from the far-reaching impacts of climate change to the ramifications of rapid urbanization and disparities in resource distribution. To realize this goal, the imperative lies in nurturing resilient food systems, propelling technological innovations, and fostering cross-border collaborations (UN Goal 2: Zero Hunger n.d.; Cohen 2017; Inter-Agency and Expert Group on Sustainable Development Goal Indicators (IAEG-SDGs) n.d.).

The interconnectedness of SDG 2 with the entire spectrum of SDGs is undeniable and bears immense importance (Wong 2021). For instance, the elimination of poverty, the enhancement of healthcare and education, and the advancement of gender equality are all integral to ensuring food security and eradicating malnutrition. Similarly, efforts to combat climate change, promote sustainable consumption and production, and enhance partnerships across sectors are all enablers of achieving SDG 2. By simultaneously addressing the multifaceted challenges within these interconnected goals, SDG 2 assumes a central role in fostering a world characterized by equity, resilience, and sustainability (Dörgő et al. 2018a; Guachalla 2023).

The SDGs provide a visionary blueprint for confronting multifarious global challenges through collective action (UN Resolution Adopted by the General Assembly on 6 July 2017 n.d.; Wong 2021). Among these, SDG 2 stands as a linchpin, intricately woven into the fabric of all other goals. By synergistically addressing poverty, health, education, inequality, climate change, and more, while championing sustainable agricultural practices and inclusive development, SDG 2 serves as a lighthouse, guiding humanity towards a more just, prosperous, and sustainable future.

2. Methodology

The information provided in this chapter draws upon an extensive analysis of both peer-reviewed literature and grey literature. A comprehensive exploration of relevant publications was conducted across renowned databases, namely Google Scholar, Scopus, Institute of Scientific Information (ISI), and the Scientific Citation Index (SCI) Web of Science. This chapter extensively considers international literature, encompassing a total of 131 peer-reviewed and grey literature studies. Of this total, three publications were authored by the authors. The time span covered by the chosen papers was open-ended; however, all the studies featured in this review maintained a focus on SDGs.

The search process employed keywords, such as "SDG 2: Zero Hunger", "SDGs", "policies and programs addressing the causes of hunger", and "shortcomings and limitations of SDGs". The ensuing dataset facilitated the identification of key themes aligned with the objectives delineated in this chapter.

Probing the intricate web of connections between SDG 2: Zero Hunger and the remaining 16 Sustainable Development Goals (SDGs) required a systematic analysis to distinguish direct and indirect links. By scrutinizing the objectives, underlying principles, and underlying interdependencies of each SDG, a comprehensive understanding of their intricate relationships emerged.

The identification of direct connections between SDG 2 and other goals was relatively straightforward. In cases where the language of the SDGs explicitly mentioned hunger, food security, or nutrition, a direct relationship was evident. For instance, the alignment between SDG 1: No Poverty and SDG 2 was palpable, as poverty often underpins hunger and malnutrition. Similarly, SDG 3: Good Health and Well-being explicitly ties nutrition to health, solidifying a direct link between these goals.

Indirect links were discerned by analyzing the synergies, shared objectives, and common principles between SDG 2 and other goals. Indicators that signaled a broader connection included themes, such as sustainable production, responsible consumption, equitable access to resources, environmental conservation, and economic growth. For example, the resonance between SDG 2 and SDG 8: Decent Work and Economic Growth was clarified by the shared emphasis on enhancing livelihoods, while the alignment between SDG 2 and SDG 12: Responsible Consumption and Production was revealed through their mutual focus on sustainable resource utilization.

In cases where multiple SDGs addressed overlapping themes, a holistic perspective was applied to ascertain the nature of the connection. For instance, while the link between SDG 2 and SDG 13: Climate Action was not overt, it emerged as a crucial interrelation when considering the impact of climate change on food systems and agricultural productivity.

Furthermore, recognizing the systemic nature of the SDGs, the interconnectedness between goals was assessed through a perspective of sustainability. Goals promoting education, gender equality, clean energy, and strong institutions were identified as underpinning the foundation for achieving SDG 2 (Otto-Zimmermann 2012). These indirect relationships became evident as the pursuit of these goals contributed to enhancing food security, sustainable agriculture, and equitable access to resources.

The identification of direct and indirect connections between SDG 2: Zero Hunger and the 16 other SDGs was a process of meticulous analysis, considering explicit mentions, shared objectives, thematic alignments, and holistic sustainability perspectives. Also, the concrete examples of good practices and innovations in the field of sustainable food and agriculture, as well as related fields, identified in this chapter, demonstrate how SDG 2: Zero Hunger can be achieved through integrated and innovative solutions.

This comprehensive approach clarified the intricate fabric of interdependencies. Analyzing the targets of each SDG separately, we identified common key elements that were considered relevant in presenting the link between SDG 2: Zero Hunger and the other SDGs.

3. Characteristics and Mutual Key Elements of Links between SDG 2: Zero Hunger and Other SDGs

3.1. The Link between SDG 1 and SDG 2

The major challenge in the link between SDG 2 and SDG 1 lies in addressing the inseparable problem of global poverty and hunger. This direct connection is crucial for resolving a complex dilemma. SDG 1 and SDG 2 are closely intertwined, and achieving their objectives can contribute to building a more sustainable, equitable, and prosperous world (Pakkan et al. 2023; UN Resolution Adopted by the General Assembly on 6 July 2017 n.d.).

There is a direct link between SDG 1: No Poverty and SDG 2: Zero Hunger. Understanding this link is important in addressing the inseparable problem of global poverty and hunger. SDG 1 and SDG 2 are directly interconnected, and achieving their goals can help achieve a more sustainable, equitable, and prosperous world (Figure 1).

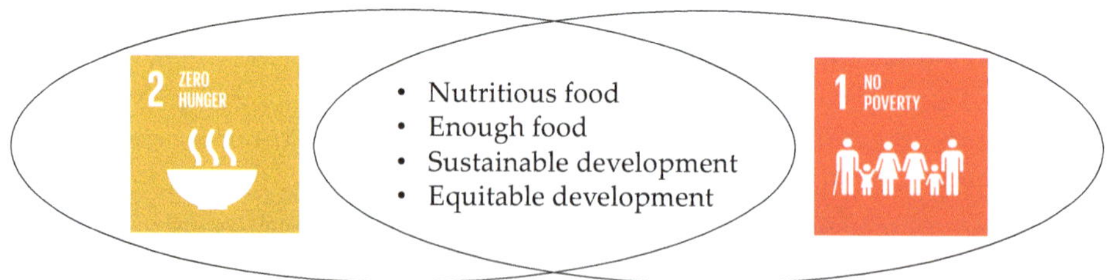

Figure 1. Mutual key elements of SDG 1 and SDG 2. Source: Figure by authors.

Poverty and hunger are often interconnected and form a vicious circle. People living in poverty are often unable to secure the necessary daily food, exacerbating malnutrition and health problems. At the same time, people suffering from malnutrition and food-related illnesses are often too sick or weak to work and support themselves, which can lead to poverty (Pakkan et al. 2023; Pradhan et al. 2017).

Therefore, reducing poverty and hunger is an interrelated and essential goal to achieve sustainable development (Box 1). Improving access to nutritious and sustainable food can help reduce poverty by increasing the productivity and incomes of farmers and others in the food sector, as well as creating jobs in the sector (Guang-Wen et al. 2023; Tremblay et al. 2020). Poverty reduction can also help increase access to food by increasing the purchasing power of the poor (Inter-Agency and Expert Group on Sustainable Development Goal Indicators (IAEG-SDGs) n.d.; UN Global SDG Indicators Database n.d.; Janoušková et al. 2018).

Box 1. Interlinked actions to reduce hunger and poverty.

- *Bangladesh: In poor rural areas of developing countries, people often make a living from agriculture and fishing. If these people do not have access to resources such as seed irrigation and proper farming tools, they cannot produce enough food to feed their families and sell the surplus to make a living. Thus, poverty makes them vulnerable to hunger and malnutrition. The Bangladeshi government has launched several poverty reduction programs including the "National Social Protection Program" welfare program and the "IFAD" program, which provide loans and training to poor farmers. Bangladesh has also made significant progress in fighting hunger by improving agricultural production, access to drinking water, and transport infrastructure (International Fund for Agricultural Development—Bangladesh n.d.).*
- *India: Poor people often live in urban slum areas where resources are limited. These people often live in unsanitary and overcrowded conditions without access to adequate food or clean water. These poor living conditions make them vulnerable to nutrition-related diseases and other conditions, such as diabetes and heart disease, which can lead to malnutrition and ultimately starvation. If poor people do not have access to education and decent jobs, they cannot afford to procure nutritious food and ensure food security for their families. In addition, poverty can lead to forced migration, which can exacerbate the problem of hunger in regions affected by conflict and natural disasters. The Government of India has launched a welfare program called "Pradhan Mantri Garib Kalyan Yojana" (PMGKY), which aims to provide financial aid to poor and vulnerable individuals affected by the COVID-19 pandemic. Additionally, India has introduced a series of agricultural policies aimed at improving food security including the "Pradhan Mantri Fasal Bima Yojana" (PMFBY) program which offers crop insurance (Pradhan Mantri Garib Kalyan Yojana (PMGKY), India n.d.).*

3.2. The Link between SDG 2 and SDG 3

The challenges in the link between SDG 2: Zero Hunger and SDG 3: Good Health and Well-being lie in the intricate interplay between adequate food, nutrition, and overall health. The essential relationship between these goals underscores that hunger and undernutrition can adversely impact health and immunity (Guang-Wen et al. 2023; Tremblay et al. 2020). The challenge emerges from orchestrating a harmonious synergy between these two goals to optimize overall health and well-being while effectively curbing hunger and malnutrition (Pakkan et al. 2023; Pradhan et al. 2017).

Adequate food and nutrition are essential for people's health and well-being, and hunger and undernutrition can adversely affect health and immunity (FAO, IFAD, UNICEF, WFP and WHO 2020).

Therefore, promoting adequate nutrition and sustainable food production can significantly contribute to improving people's health and well-being and achieving SDG 3 (Figure 2).

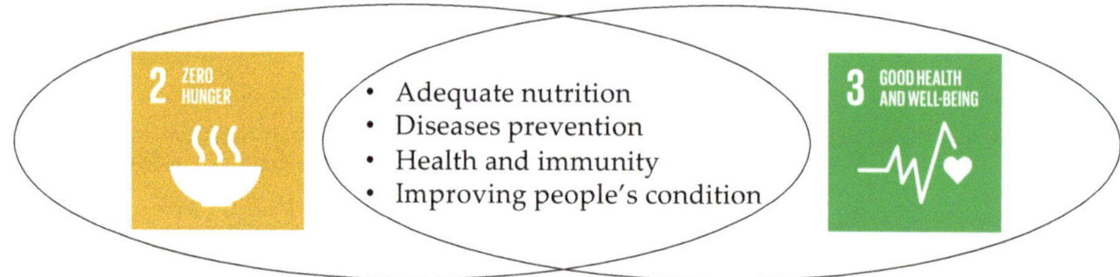

Figure 2. Mutual key elements of SDG 2 and SDG 3. Source: Figure by authors.

Achieving SDG 2 and SDG 3 is closely and directly linked, as adequate food and nutrition are essential for health and well-being, and hunger and undernutrition can adversely affect health.

SDG 3 aims to ensure universal access to health services and promote physical and mental health, and adequate nutrition plays an important role in achieving this goal. Nutritious and balanced foods provide the body with the nutrients it needs to prevent diseases associated with poor nutrition, such as malnutrition, obesity, and diabetes (FAO, IFAD, UNICEF, WFP and WHO 2020).

On the other hand, malnutrition and undernutrition can affect a person's immune system, making them more vulnerable to disease and infection. By ensuring access to nutritious and balanced food, you can help boost immunity and prevent disease.

SDG 2 aims to reduce the number of people suffering from hunger and malnutrition by promoting access to nutritious and sustainable food. In addition, SDG 2 aims to improve food production by using sustainable agricultural practices and by promoting an efficient and responsible food system (Box 2).

Box 2. Interlinked actions to reduce hunger and to increase health and well-being.

- *Brazil: National School Feeding Program: This program provides nutritious food to approximately 42 million students in public schools in Brazil, thereby promoting adequate nutrition and combating malnutrition. Brazil has also introduced a law banning the sale of unhealthy food in public schools (FAO 2015).*
- *India: National Nutrition Mission: This program aims to improve the nutrition and health of children aged 0–6 years. The program provides nutritional supplements and nutrition education to mothers and children in poor rural and urban communities (National Nutrition Mission, India n.d.).*
- *USA: SNAP Program: The Supplemental Nutrition Assistance Program (SNAP) provides food assistance to millions of low-income Americans who are experiencing hunger or malnutrition. The program also encourages the purchase of nutritious food and promotes access to local agricultural markets (Supplemental Nutrition Assistance Program (SNAP) USA n.d.).*
- *USA: Farm-to-School Initiative: This initiative promotes the purchase of local and sustainable food products for schools in Delaware State, thereby helping to reduce environmental impacts and promote healthy eating (Farm-to-School USA n.d.).*

3.3. The Link between SDG 2 and SGD 4

There is an indirect link between SDG 2: Zero Hunger and SDG 4: Quality Education. The challenges in the connection between SDG 2: Zero Hunger and SDG 4: Quality Education arise from the intricate indirect relationship between these goals. While promoting access to quality education can indirectly contribute to achieving SDG 2 by enhancing knowledge about healthy eating, food production, and sustainable practices, the effectiveness of this link relies on several factors. Education's potential to empower individuals to make informed choices regarding food and support sustainable agriculture practices faces challenges in ensuring widespread access to quality education (Pradhan et al. 2017). The alignment between SDG 2 and SDG 4 depends on addressing barriers to education, such as limited access, gender disparities, and economic constraints.

While education can equip people with practical skills related to food cultivation, preparation, and sustainable farming techniques, translating these skills into actionable steps to reduce hunger necessitates overcoming barriers, like limited resources and infrastructure. Moreover, promoting gender equality through education can positively influence access to food, but this requires addressing deeply ingrained societal norms and overcoming gender-related barriers to education (Guang-Wen et al. 2023; Balan et al. 2022a; Tremblay et al. 2020). Similarly, while education's role in reducing poverty has the potential to enhance access to food, overcoming systemic economic challenges and ensuring equitable educational opportunities are essential components of bridging the gap between these goals. Thus, while the link between SDG 2 and SDG 4 holds potential, the challenges lie in addressing disparities in education access, gender equality, and poverty reduction to effectively contribute to hunger alleviation and sustainable food practices (Di Fabio and Rosen 2020; Pradhan et al. 2017).

Promoting access to quality education can contribute to achieving SDG 2 and reducing hunger in several ways. Education can help people understand the importance of healthy eating and how food is grown and processed. Thus, people can make informed choices when it comes to food and support sustainable food and agricultural practices that can help achieve SDG 2 (Figure 3).

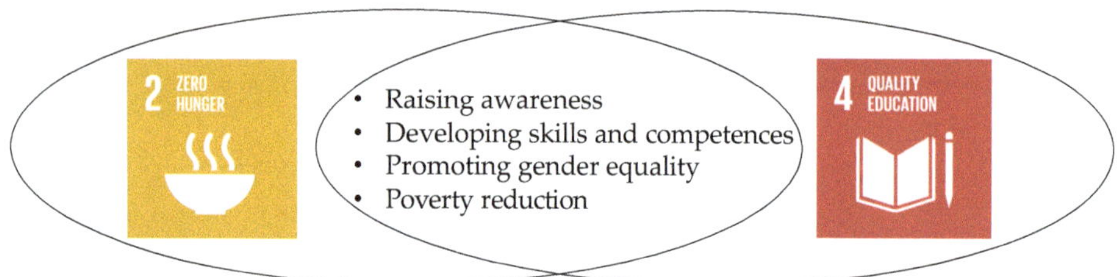

Figure 3. Mutual key elements of SDG 2 and SDG 4. Source: Figure by authors.

Promoting access to quality education can significantly contribute to achieving SDG 2 and reducing hunger by raising awareness, developing skills and competences, promoting gender equality, and reducing poverty. Education can help people learn practical skills, such as growing vegetables and fruit, preparing nutritious food, and using sustainable farming techniques. These skills can help people sustainably feed their families and communities and help achieve SDG 2. Education can help promote gender equality, which can have a positive impact on access to food. In many communities, women grow food and prepare food, and education can help increase their chances of accessing resources and becoming leaders in their community (Pakkan et al. 2023).

Education can help reduce poverty, which can have a direct impact on access to food. People who have access to education are more likely to find a well-paying job and improve their financial situation, which can enable them to buy more nutritious food and feed their families in a more sustainable way (Box 3).

Box 3. Interlinked actions to reduce hunger and to increase quality education.

- *Ghana—The Ghana National School Feeding Program was launched in 2005 and aims to provide free nutritious food to primary school students. Through this program, Ghana has been able to improve the quality of food consumed by students and encourage parents to send their children to school. In addition, the program has had a positive impact on the local economy, as food is purchased from local producers. In addition, as well as academic subjects, children are also taught about growing plants, food processing, and nutrition. Thus, children are taught how to grow their own food and prepare healthy and nutritious meals, which helps to raise awareness and develop the skills and competencies needed to help achieve SDG 2 (Ghana School Feeding Programme (GSFP) n.d.).*
- *Brazil—The Family Agriculture Food Acquisition Program (PAA) in Brazil was launched in 2003 and aims to promote local agriculture and provide healthy food for public schools in rural areas. Through this program, schools buy food directly from local producers, which helps them sell their products and improve their income. In addition, students receive healthy and nutritious food, which helps reduce hunger and improve their health. An educational program was launched for children from low-income families. In the program, children learn about healthy eating, how food is processed, and the impact of food on health and the environment. Also, children are taught how to cook healthy and nutritious meals with local and sustainable food. Thus, children learn to choose healthy foods and cook nutritious meals, which helps to raise awareness and develop the skills and competencies needed to help achieve SDG 2 (De Souza et al. 2023).*
- *India—India's National Mid-Day Meal School Feeding Program was launched in 1995 and aims to provide free nutritious food to primary and secondary school students in rural and urban areas. Through this program, India has been able to increase school attendance and improve children's nutrition. In addition, the program has had a positive impact on the local economy, as food is purchased from local producers. At the same time, a professional training program for women was established. In the program, women are taught how to grow vegetables and fruits in a sustainable way and how to process and sell their produce. Also, women learn about nutrition and how food can be used to fight hunger and malnutrition in their community. Thus, the program contributes to the development of skills and competences needed to help achieve SDG 2, by promoting gender equality and reducing poverty (Mishra 2023).*

3.4. The Link between SDG 2 and SGD 5

There is an indirect link between SDG 2: Zero Hunger and SDG 5: Gender Equality. The challenges in the connection between SDG 2: Zero Hunger and SDG 5: Gender Equality stem from the intricate indirect relationship that relies on addressing gender disparities and empowering women. While gender equality is essential for achieving SDG 2: Zero Hunger, challenges arise in transforming this linkage into tangible outcomes.

Women's pivotal role in agriculture and food production, coupled with limited access to resources and technology, presents a challenge in bridging the gap between gender equality and hunger alleviation (Hoddinott and Haddad 1995). While promoting gender equality can facilitate women's access to resources and technology for improved agricultural activities and food production, addressing deeply rooted gender norms and overcoming systemic barriers to women's empowerment are critical. The vulnerability of women to hunger and malnutrition due to limited access to food, healthcare, and gender discrimination underscores the need for gender-sensitive strategies. However, translating gender equality initiatives into reduced hunger demands navigating the complexity of societal perceptions and deeply ingrained biases. Ensuring women's participation in decision-making processes is vital, yet challenges lie in dismantling existing power structures and fostering an inclusive environment (FAO, IFAD, UNICEF, WFP and WHO 2020).

Promoting gender equality holds the potential to significantly contribute to SDG 2 by enhancing food production and reducing hunger and malnutrition. Nevertheless, challenges revolve around confronting traditional norms and biases that hinder women's access to resources, healthcare, education, and decision-making processes (Di Fabio and Rosen 2020). Achieving true gender equality requires addressing deeply rooted systemic issues and enacting comprehensive policies that empower women economically, socially, and politically. Consequently, the success of the link between SDG 2 and SDG 5 lies in dismantling barriers that restrict women's potential, fostering inclusivity, and creating an enabling environment that transcends stereotypes and biases to effectively reduce hunger and improve food security (Lucato et al. 2018).

Gender equality is important for achieving SDG 2 for several reasons (Figure 4).

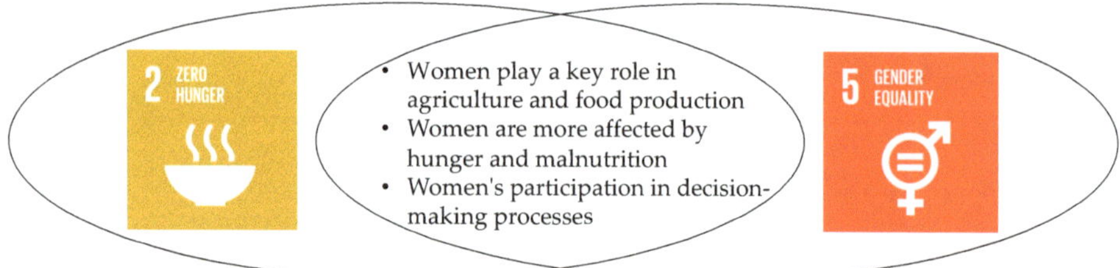

Figure 4. Mutual key elements of SDG 2 and SDG 5. Source: Figure by authors.

Women play a key role in agriculture and food production because, in many countries, women are responsible for agricultural work and food production. However, they often have limited access to the resources needed to develop their agricultural activities and improve their food production. Promoting gender equality can ensure that women have access to the resources and technology needed to develop their agricultural activities and improve their food production. This can contribute to achieving SDG 2 by increasing food production and reducing hunger.

Women are more affected by hunger and malnutrition, because women are often more vulnerable to hunger and malnutrition than men. This is due to several factors, such as limited access to food and medical services, as well as gender discrimination. Promoting gender equality can ensure that women have access to food and health services, as well as equal opportunities in education and the labor market. This can contribute to achieving SDG 2 by reducing hunger and malnutrition.

Women's participation in decision-making processes is very important, because promoting gender equality can ensure that women have a greater role in decision-making processes related to agricultural policy and food production. This can help create more effective policies to reduce hunger and malnutrition (Guang-Wen et al. 2023; Tremblay et al. 2020).

Promoting gender equality can significantly contribute to achieving SDG 2 by increasing food production and reducing hunger and malnutrition (Box 4). Ensuring access to the necessary resources and technology for women, as well as promoting equal participation of women in decision-making processes, can contribute to the development of more effective policies and programs to reduce hunger and undernutrition (Pakkan et al. 2023).

Box 4. Interlinked actions to reduce hunger and to increase gender equality.

- *In Rwanda, the "One Cow per Family" program provides cows to women in poor communities to increase milk and meat production, which has led to improved family nutrition and increased incomes. Under this program, women are given priority in receiving the cows, as they are considered to need more resources to support their families. Thus, promoting gender equality and ensuring women's access to agricultural resources can contribute to achieving SDG 2 (Nilsson et al. 2019).*
- *In India, the Organization of Women Honey Producers (WIHPA) helps women produce and market honey by providing them with access to equipment, funding, and training in honey production. These agricultural activities are an important source of income for women and help reduce poverty and hunger. Promoting gender equality in the agricultural sector can increase women's access to such opportunities and could contribute to achieving SDG 2 (Organization of Women Honey Producers (WIHPA), India n.d.).*
- *In Nepal, the Maternal and Child Health and Nutrition Program (WHCNP) aims to reduce maternal and child malnutrition through education and health services. Within the program, special attention is paid to promoting gender equality through mothers' groups and counseling sessions. Improving the health of women and children can help reduce undernutrition and achieve SDG 2 (Maternal, Child Health and Nutrition Project, Nepal n.d.).*

3.5. The Link between SDG 2 and SGD 6

There is an indirect link between SDG 2: Zero Hunger and SDG 6: Clean water and sanitation. The challenges in the connection between SDG 2: Zero Hunger and SDG 6: Clean Water and Sanitation arise from the intricate interplay between food security and water access. While the indirect link between these goals highlights their mutual influence, challenges emerge in ensuring holistic solutions to food and water-related issues. Recognizing the role of clean water and proper hygiene in food production underscores the importance of access to clean water in enhancing agricultural productivity (Dora et al. 2015; Hutton and Haller 2004). Yet, challenges exist in guaranteeing sufficient and safe water sources for irrigation, especially in regions where water scarcity prevails. Addressing this challenge necessitates innovative water management strategies and technologies that can sustainably support agriculture.

Furthermore, the interrelation between access to clean water and foodborne illnesses poses a challenge in reducing the risks of malnutrition and disease transmission. Ensuring clean water sources for agriculture and proper hygiene practices is crucial to prevent the contamination of food and waterborne diseases. However, overcoming infrastructural and awareness-related hurdles to ensure the availability of clean water in rural areas and promoting hygienic practices remains a challenge in many regions (Pakkan et al. 2023).

The prevention of conflicts over water resources is another challenge embedded in the linkage between SDG 2 and SDG 6. As water scarcity intensifies globally, competition for water resources becomes more pronounced, particularly in regions where water availability is limited. While clean water is essential for food production, challenges emerge in maintaining equitable access to water resources for both agriculture and domestic use (Simonovic 2002). Collaborative efforts are needed to establish effective water management frameworks that promote sustainability and prevent disputes over water allocation.

The success of the connection between SDG 2 and SDG 6 hinges on addressing challenges associated with water scarcity, contamination, and equitable water resource distribution. Collaborative approaches that involve governments, local communities, and international organizations are crucial for devising and implementing water management strategies that ensure clean water access for agriculture, promote proper hygiene practices, and prevent conflicts over water resources. This demands a comprehensive and integrated approach that recognizes the intricate relationship between food security and water availability, while navigating challenges specific to each region's context.

Understanding this link is important for addressing food and water security issues in an integrated manner (Figure 5).

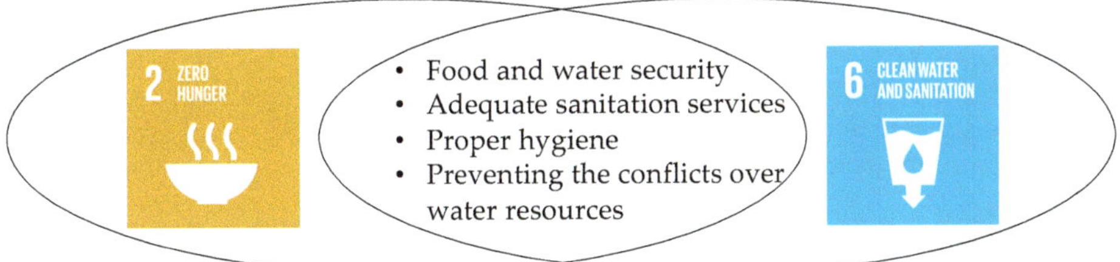

Figure 5. Mutual key elements of SDG 2 and SDG 6. Source: Figure by authors.

Clean water and hygiene are essential for food production. This is because water is one of the most important factors for agriculture and food production. Farmers need water to irrigate crops and feed animals. Ensuring access to clean water and adequate sanitation services can help increase food production and reduce hunger (Box **??**).

Access to clean water and proper hygiene can help prevent foodborne illness. Contaminated water can contribute to the transmission of foodborne illness and increase the risk of malnutrition. By promoting access to clean water and proper hygiene, the risk of disease and malnutrition can be reduced.

This can help prevent conflicts over water resources. This is especially important in areas where water resources are limited. Access to clean water and adequate sanitation services can help prevent conflicts and promote the sustainable use of water resources.

Box 5. Interlinked actions to reduce hunger and to increase access to clean water and sanitation.

- *In rural Africa, many farmers rely on subsistence agriculture, and water is a crucial factor in this activity. Providing access to drinking water sources and irrigation technologies can support increased agricultural production and food security in poor regions of Africa (Mbatha et al. 2021).*
- *In some parts of Nigeria, children are missing out on going to school because of water-related diseases. For example, diarrhea and water-borne infections can prevent children from learning and developing intellectual skills. By providing clean water and adequate sanitation services, disease can be reduced and, thus, education and economic development can be supported in these communities (Ali 2022).*

- *In many parts of the world, women and girls are responsible for collecting water. In rural Africa, this task can take up several hours a day, time that they cannot spend on productive activities such as farming or education. Providing drinking water sources close to communities can help reduce this burden and, thus, allow women and girls to devote more time to productive activities (Kayser et al. 2019; Parry and Gordon 2021).*
- *In congested cities where water is limited, domestic water can be reused for plant irrigation and other agricultural activities. Thus, reducing household water use can support reducing water use for agriculture and increasing water resources available for both purposes. This can help reduce hunger and malnutrition in urban areas, as well as promoting the sustainable use of water resources (Chen et al. 2021; Egbuikwem et al. 2020).*

3.6. The Link between SDG 2 and SGD 7

There is an important indirect link between SDG 2: Zero Hunger and SDG 7: Clean and Affordable Energy. This link mainly concerns the need to use energy sustainably to support food production and improve access to food sustainably.

The challenges embedded in the connection between SDG 2: Zero Hunger and SDG 7: Clean and Affordable Energy revolve around the intricate balance between energy sustainability and food security. While the indirect link highlights the significance of utilizing energy resources efficiently to support food production and distribution, several challenges need to be addressed to achieve these objectives sustainably. The requirement for energy in various stages of the food supply chain, from cultivation and processing to transportation and storage, underscores the importance of accessing clean and affordable energy sources. However, challenges emerge in promoting the adoption of sustainable energy solutions, such as solar or wind energy, across all sectors of the food system. Overcoming financial and infrastructural barriers to integrating renewable energy technologies into food production processes remains a challenge in many regions.

Moreover, ensuring equitable access to clean and affordable energy resources is crucial for improving access to food sustainably. Challenges arise in remote and underserved areas, where energy infrastructure is lacking, hindering efficient food storage and preservation, as well as agricultural production processes (Singh et al. 2018). Addressing these challenges requires collaborative efforts between governments, the private sector, and international organizations to develop and implement energy infrastructure projects tailored to the specific needs of these communities.

The link between SDG 2 and SDG 7 also encompasses the imperative of reducing food loss. While access to clean and affordable energy can enhance food storage and transportation systems, challenges exist in ensuring the widespread adoption of energy-efficient technologies that minimize food loss and waste. This demands investment in innovative solutions and awareness-raising initiatives to promote sustainable practices across the food supply chain (Pakkan et al. 2023).

Ultimately, addressing the challenges within the connection between SDG 2 and SDG 7 requires a holistic approach that accounts for energy sustainability, food security, and equitable access. Collaborative endeavors are needed to foster technological innovation, policy development, and financial support for clean and affordable energy solutions, enabling sustainable food production, reducing food loss, and enhancing access to nutritious food (Singh et al. 2018; Di Fabio and Rosen 2020). This intricate relationship underscores the need for integrated strategies that acknowledge the potential of sustainable energy to contribute to achieving SDG 2, while also considering the diverse challenges specific to various regions and contexts (Figure 6).

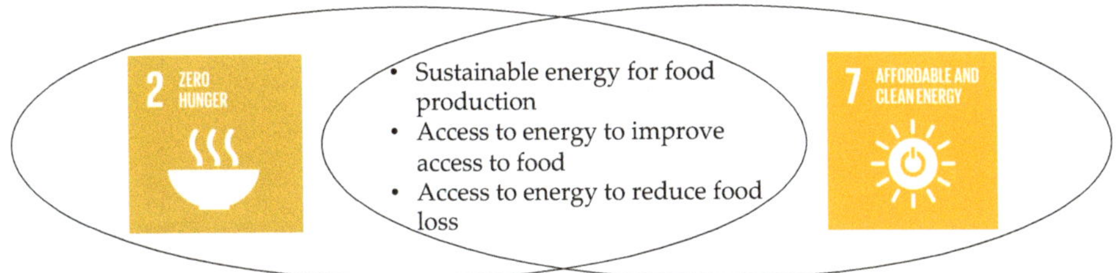

Figure 6. Mutual key elements of SDG 2 and SDG 7. Source: Figure by authors.

Food production requires energy, for example, to irrigate crops, to process and package food, or to transport it to consumers. Using energy from sustainable sources, such as solar or wind energy, can reduce greenhouse gas emissions and help ensure sustainable food production.

Access to clean and affordable energy can improve access to food by improving food storage and preservation, by increasing the efficiency of production processes and by improving access to refrigeration systems in rural and peripheral areas (Singh et al. 2018). Access to energy can help reduce food loss by improving the efficiency of refrigeration systems, by improving transportation systems, and by improving access to food storage and preservation systems.

So, the use of energy from sustainable sources can help ensure sustainable food production and reduce food loss, which can help achieve SDG 2. Also, access to clean and affordable energy can improve access to food and can help reduce hunger and malnutrition (Box 6).

Box 6. Interlinked actions to reduce hunger and to increase access to clean and affordable energy.

- *Using solar energy to irrigate crops—In regions where irrigation is needed to increase food production, solar energy can be used to power the necessary water pumps. This can be a sustainable solution that reduces dependence on unsustainable energy sources and can help increase food production (Arshad and Khalid 2022; Schnetzer and Pluschke 2018).*
- *Using renewable energy for food processing—Energy is needed to process and package food. The use of renewable energy, such as wind power or hydroelectric power, can reduce greenhouse gas emissions and contribute to more sustainable food production (Bielski et al. 2021; Anser et al. 2021).*
- *Using clean energy to transport food—Transporting food to consumers requires energy. The use of fossil fuels leads to greenhouse gas emissions and climate change. Using clean energy, such as electricity, can help reduce environmental impacts and increase the sustainability of food production (Sharma et al. 2021; Jouzdani and Govindan 2021).*
- *Using clean energy for food storage and preservation—Access to energy can help improve food storage and preservation. Refrigeration systems are necessary to keep food fresh and safe for consumption. The use of clean energy can help reduce energy costs and increase the affordability of refrigeration systems in rural and peripheral areas (Pandiselvam et al. 2019).*

3.7. The Link between SDG 2 and SGD 8

There is an indirect link between SDG 2: Zero Hunger and SDG 8: Decent Work and Economic Growth. This link refers to the fact that reducing poverty and hunger can contribute to economic growth and the creation of decent jobs. The agricultural sector plays an important role in the global economy, and increasing productivity and market access for farmers in rural areas can contribute to economic growth and the creation of decent jobs (Guang-Wen et al. 2023).

The challenges inherent in the relationship between SDG 2: Zero Hunger and SDG 8: Decent Work and Economic Growth encompass intricate socio-economic dynamics that require careful consideration. The indirect link between these two goals underscores the potential of reducing poverty and hunger to catalyze economic growth and the emergence of decent employment opportunities. While the connection emphasizes the role of the agricultural sector in this process, a range of challenges must be addressed to realize this potential comprehensively (Tremblay et al. 2020).

Agriculture's pivotal role in the global economy highlights the significance of enhancing productivity and market access for rural farmers. Yet, challenges persist in ensuring the equitable distribution of resources, technology, and training, particularly in marginalized regions. Overcoming barriers to accessing markets,

3.9. The Link between SDG 2 and SGD 10

There is an indirect link between SDG 2: Zero Hunger and SDG 10: Reducing Inequalities. Poverty and inequality are major contributors to hunger and malnutrition. At the same time, hunger and malnutrition can exacerbate inequality and poverty.

The interconnectedness between SDG 2: Zero Hunger and SDG 10: Reducing Inequalities entails multifaceted challenges that underscore the intricate relationship between hunger, poverty, and inequality. While an indirect link exists between these goals, their symbiotic nature underscores the imperative of addressing them in tandem. Hunger and malnutrition are both consequences and causes of inequality and poverty, disproportionately impacting marginalized groups such as those residing in rural areas or extreme deprivation. Resolving this predicament necessitates concerted efforts to address challenges tied to access, resources, and living standards for disadvantaged communities (UN Progress towards the Sustainable Development Goals n.d.).

Inequalities can engender food inaccessibility, rendering it unaffordable for impoverished populations who lack the means to secure nutritious meals or cultivate their sustenance due to limited resources and land access (FAO, IFAD, UNICEF, WFP and WHO 2021). Reducing inequalities holds the key to mitigating hunger and malnutrition by ensuring equitable access to food and essential health services for all, regardless of socioeconomic status. Correspondingly, combating hunger can be a catalyst for reducing inequality, uplifting living conditions, and enhancing resource availability for marginalized communities.

Harmonizing efforts to alleviate hunger, malnutrition, and inequality is pivotal to fostering a society where all individuals can access nourishing sustenance and a decent standard of living. The convergence of SDG 2 and SDG 10 objectives signifies a collaborative approach that leverages policy interventions, resource allocation, and community engagement to effect transformative change. Empowering marginalized communities through equitable access to food, health services, education, and opportunities is a fundamental step toward realizing the vision of eradicating hunger and inequalities (FAO, IFAD, UNICEF, WFP and WHO 2021). Thus, intertwining the pursuit of SDG 2 and SDG 10 yields a synergistic approach that resonates with the essence of sustainable development (Figure 9).

Hunger and malnutrition disproportionately affect disadvantaged communities, such as those in rural areas or those living in extreme poverty. These communities need support and investment to gain access to food and improve their living standards.

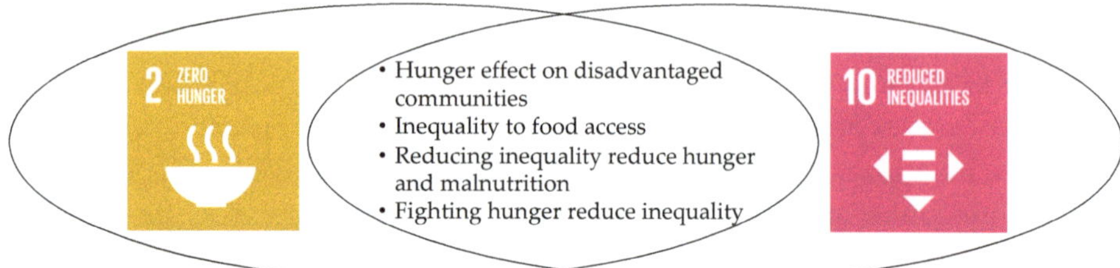

Figure 9. Mutual key elements of SDG 2 and SDG 10. Source: Figure by authors.

Inequality can make food unaffordable for some people, especially those living in poverty. They cannot afford to buy nutritious food or grow their own food due to lack of resources and access to land (FAO, IFAD, UNICEF, WFP and WHO 2020).

Reducing inequality can help reduce hunger and malnutrition by improving access to food and health services for all people, regardless of social position or income. Fighting hunger can help reduce inequality by improving living standards and access to resources for disadvantaged communities (Box 9).

Box 9. Interlinked actions to reduce hunger and inequalities.

- *India—In India, there is a wide disparity between rural and urban areas in terms of access to nutritious food. In recent years, the Indian government has launched food distribution programs for the poorest citizens, as well as projects to improve infrastructure in rural areas to help people grow their own food. These efforts have had a positive impact on reducing hunger and malnutrition in the country and have contributed to a reduction in inequalities (George and McKay 2019).*
- *Brazil—In Brazil, the government introduced a program called "Fome Zero" (Zero Hunger), which aims to eliminate hunger and malnutrition in the country. The program includes distributing free food to poor families, improving access to drinking water, and promoting sustainable agriculture. These efforts have had a significant impact on reducing hunger and malnutrition in Brazil, particularly among disadvantaged communities (Da Silva et al. 2011).*
- *Kenya—In Kenya, the government has launched a program called "Kilimo Biashara" (Agriculture for Business), which aims to support local agriculture and promote food security in the country. The program includes the distribution of seeds and agricultural machinery, as well as training farmers in modern farming techniques. These efforts have helped improve food production and reduce hunger and malnutrition in Kenya (Kilimo Biashara Program, Kenya n.d.).*

3.10. The Link between SDG 2 and SGD 11

SDG 11: Sustainable cities and communities is also directly linked to SDG 2: Zero Hunger. The goal of SDG 11 is to make cities and human settlements more inclusive, safer, more resilient, and more sustainable. Urbanization and the growth of cities have an impact on food systems and agriculture. As cities expand, the agricultural space around them shrinks, which can lead to a decrease in local food production and an increase in dependence on imports. Urbanization can also lead to increased consumption of processed and fast food, which can be less healthy and less nutritious than fresh, local food.

The interconnection between SDG 2: Zero Hunger and SDG 11: Sustainable Cities and Communities presents a complex tapestry of challenges emanating from the dynamic relationship between urbanization, food systems, and sustainable development. The symbiotic bond between these goals accentuates the intricate interplay between urban growth and food security. Urbanization's expansion and city development hold implications for agriculture and food systems, often leading to diminished local food production due to shrinking agricultural spaces and escalating reliance on imports. Concurrently, urbanization fosters an upsurge in the consumption of processed and fast foods, potentially compromising nutritional quality (FAO, IFAD, UNICEF, WFP and WHO 2020).

The nexus of SDG 2 and SDG 11 centers on ensuring urban communities' access to healthy, safe, and nutritious food. In this vein, championing urban agriculture and fortifying food storage and distribution infrastructure are pivotal shared aspirations. Moreover, both goals underscore the imperative of judiciously managing natural resources. Urban agriculture can counterbalance food imports, optimize land use, and bolster local economies by generating jobs in agriculture and associated sectors. This aligns with SDG 2 and SDG 11 objectives of fostering economic growth and sustainable development.

Furthermore, the objectives of sustainable urbanization and appropriate resource management are underscored by urban agriculture's potential to curtail greenhouse gas emissions via reduced food transport and enhanced energy efficiency in farming practices. Yet, the intrinsic linkage between SDG 2 and SDG 11 confronts challenges, particularly the negative impact of unchecked urban development on soil, water, and air quality that can undermine agricultural productivity and food quality (Nilsson et al. 2016).

Thus, the link between SDG 2 and SDG 11 resonates with joint actions to promote resilient, sustainable, and inclusive communities. A holistic urban development strategy encompassing improved infrastructure, transport systems, and responsible urban planning can facilitate food access and distribution. To harmonize these goals, initiatives must address both urban challenges and agricultural resilience, thereby forging pathways toward zero hunger and sustainable urban progress (FAO, IFAD, UNICEF, WFP and WHO 2020). In essence, the intertwined narratives of

SDG 2 and SDG 11 highlight the confluence of urban transformation and food security, underscoring the need for comprehensive strategies that nurture both thriving communities and nourishing sustenance (Figure 10).

Food security and access to healthy and nutritious food are essential to ensure the health and well-being of urban communities. In many cities around the world, access to fresh, healthy food is limited, and this can lead to health problems, such as obesity, diabetes, and cardiovascular disease. Cities also produce a significant amount of food waste and contribute to climate change through greenhouse gas emissions from transportation and the production of processed foods. In this context, urbanization can lead to the increased consumption of processed food and fast food, which can be less healthy and less nutritious than fresh and local food, so SDG 2 and SDG 11 are closely related and have several common elements.

Figure 10. Mutual key elements of SDG 2 and SDG 11. Source: Figure by authors.

People in urban environments need access to healthy and safe food, and this is a common goal for both SDG 2 and SDG 11. To achieve this goal, it is important to promote urban agriculture and create adequate infrastructure for food storage and distribution.

SDG 2 and SDG 11 also focus on the sustainable use of natural resources. Urban agriculture can help reduce the dependence on imported food and promote efficient land use in urban areas. SDG 2 and SDG 11 aim to create job opportunities for local communities. Urban agriculture can play an important role in creating new jobs in the agricultural sector and related services. SDG 2 and SDG 11 aim to promote sustainable development and appropriate resource management. Urban agriculture can help reduce greenhouse gas emissions by reducing food transport and increasing energy efficiency in agricultural processes.

Also, improving urban infrastructure, including transport systems, can help facilitate access to food markets and transport food from rural areas to urban areas (FAO, IFAD, UNICEF, WFP and WHO 2020).

On the other hand, uncontrolled urban development and pollution can have a negative impact on soil, water, and air quality, which can lead to a decrease in agricultural production and food quality.

There is, therefore, a strong interdependence between SDG 2 and SDG 11, as the sustainable development of cities and urban communities can significantly contribute to reducing hunger and improving food security. At the same time, uncontrolled urban development can lead to a decrease in agricultural production and food quality, which can exacerbate food security problems. Therefore, addressing the issues of sustainable development of cities and urban communities should also include solutions to support the development of the agricultural sector and to improve access to food and food security, so that it can achieve the goal of SDG 2: Zero Hunger and encourage sustainable urban development (Box 10).

Box 10. Interlinked actions to reduce hunger and increase the levels of sustainable cities and communities.

- *India: Cities in India are experiencing rapid population growth and uncontrolled urbanization, which has resulted in increased air, water, and soil pollution. This has had a significant impact on food security, as many agricultural crops are affected by pollution. In addition, access to fresh and healthy food is limited in many cities, which has led to health problems such as obesity and diabetes. To address these issues, the Indian government has launched the Smart Cities Mission, which aims to develop sustainable and inclusive cities, with a focus on improving urban infrastructure and transport. In addition, the program aims to improve food quality and food safety by supporting local agricultural production and promoting urban agriculture (Smart Cities Mission India n.d.).*
- *Brazil: Brazil is one of the largest agricultural countries in the world but also one of the most urbanized. The growth of cities led to a reduction in agricultural land and an increase in food imports. To address these issues, the Brazilian government launched the "Cidades Sustentáveis" (Sustainable Cities) program, which aims to improve the quality of life in cities through sustainable development. The program encourages the development of urban agriculture and community gardens to improve access to fresh and local food and reduce dependence on imports (Index of Sustainable Development of Cities—Brazil n.d.).*
- *United States of America: In the United States, many cities face problems of poverty and limited access to healthy and fresh food, known as "food deserts." To address these issues, the US government launched the Good Food Purchasing Program, which aims to improve food quality and food safety by promoting the purchase of local, sustainable, and healthy food in public food systems, such as schools, hospitals, and prisons (Good Food Purchasing Program USA n.d.).*

3.11. The Link between SDG 2 and SGD 12

SDG 2: Zero Hunger and SDG 12: Responsible Consumption and Production are directly linked in several ways due to their impact on the environment and natural resources. Sustainable agriculture and responsible production are central to both SDG 2 and SDG 12 (FAO, IFAD, UNICEF, WFP and WHO 2020).

The interrelation between SDG 2: Zero Hunger and SDG 12: Responsible Consumption and Production spawns an intricate web of challenges stemming from their shared commitment to environmental sustainability and resource management. The convergence of sustainable agriculture and responsible production echoes in both goals, with sustainable agricultural practices seeking to optimize food production while minimizing environmental repercussions (Salasan and Balan 2022). This duality champions SDG 2 by enhancing food security while safeguarding precious natural resources. Similarly, SDG 12's emphasis on eco-friendly production, reduced emissions, and waste prevention resonates across both goals, advocating for a greener footprint (Balan et al. 2022a).

Promoting prudent natural resource management resonates as a crucial facet in both SDG 2 and SDG 12. This encompasses judicious stewardship of agricultural land, water, and other resources, alongside curbing food and resource wastage. Sustainable resource use can bolster nutritional security, sustainable development, and environmental conservation by curtailing deforestation, preserving biodiversity, and mitigating greenhouse gas emissions (FAO, IFAD, UNICEF, WFP and WHO 2020).

Moreover, SDG 12 champions the ethos of sustainable consumption and production, underscored by a drive to minimize food waste. This facet synergizes with SDG 2 by optimizing resource utilization. Enabling a circular economy stands as a shared objective, seeking to curb natural resource consumption through enhanced recycling, reuse, and regeneration. This ethos not only addresses food waste but also elevates resource efficiency, thereby fostering economic growth and employment prospects.

Furthermore, access to clean and sustainable energy forms an instrumental nexus within SDG 2 and SDG 12. The deployment of such energy sources plays a pivotal role in curbing greenhouse gas emissions, averting climate change's detrimental impact on agriculture and food security. Moreover, clean energy catalyzes economic growth, poverty alleviation, and sustainable infrastructure development, thus harmonizing the dual aspirations of both goals.

In summation, the direct link between SDG 2 and SDG 12 navigates the trajectory towards comprehensive sustainability. Nurturing sustainable agriculture, minimizing waste, ensuring food safety, and fostering international collaboration form the keystones to realizing the dual aspirations of these SDGs. This symbiotic relationship underscores the interdependence between eradicating hunger and advancing responsible consumption and production practices, constituting a dynamic blueprint for a harmonious coexistence of humanity and the environment (Figure 11).

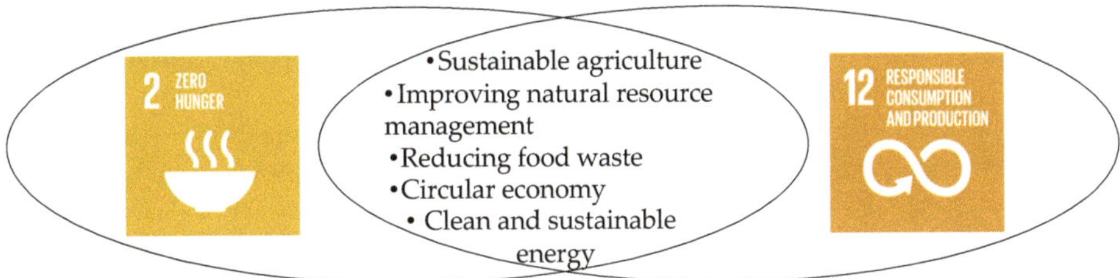

Figure 11. Mutual key elements of SDG 2 and SDG 12. Source: Figure by authors.

Sustainable agriculture promotes ecological and resource-efficient agricultural practices that can improve food production while reducing the environmental impact. This can contribute to achieving SDG 2 by increasing food security and nutrition while protecting natural resources. At the same time, responsible production aims at reducing greenhouse gas emissions, responsible use of resources, and waste prevention, all of which have a significant impact on the environment (Nilsson et al. 2016).

Improving natural resource management is another important element of SDG 2 and SDG 12. This relates to the sustainable management of agricultural land, water, and other natural resources, as well as reducing the waste of food and other resources. Sustainable use of resources can support both food and nutrition security and sustainable development by preventing deforestation, conserving biodiversity, and reducing greenhouse gas emissions.

SDG 12 encourages the development of sustainable consumption and production patterns, including reducing food waste. This can help meet SDG 2 by making more efficient use of available food resources (Balan et al. 2022b).

Promoting the circular economy is also important for both objectives. The circular economy aims to reduce the consumption of natural resources by increasing recycling, reuse, and regeneration. This can help reduce food waste and increase resource efficiency, while supporting economic growth and job creation (Nilsson et al. 2016).

Access to clean and sustainable energy is another important element of SDG 2 and SDG 12. The use of clean and sustainable energy can help reduce greenhouse gas emissions, thus preventing climate change and its impact on agriculture and food security (Box 11). In addition, clean and sustainable energy can support economic development and poverty reduction through job creation and sustainable infrastructure development (Nilsson et al. 2016).

Box 11. Interlinked actions to reduce hunger and increase responsible consumption and production.

- *Brazil—Brazil implemented the Programa de Aquisição de Alimentos (Food Acquisition Program), which encourages local and sustainable food production while ensuring access to fresh and nutritious food for low-income people (De Souza et al. 2023).*
- *Kenya—The Kenyan government has launched the Nutrition in Agriculture Program, which aims to improve agricultural productivity using sustainable methods, as well as promoting nutrition and food security for the rural population (Nutrition in Agriculture Program, Kenya—Nutrition Portal n.d.).*
- *Netherlands—The Netherlands has developed a national circular economy strategy, which aims to reduce food waste and use resources more efficiently, including water and soil. This involves, among other things, the recycling of food waste and the use of renewable energy (Government of the Netherlands. Circular Dutch Economy by 2050 n.d.).*
- *Australia—In Australia, the National Food Waste Reduction and Food Waste Phobia Mitigation Program promotes food waste reduction through education and public awareness, as well as engaging companies and organizations in developing more sustainable practices in terms of food production and distribution (Australian Government. National Food Waste Strategy 2017).*

3.12. The Link between SDG 2 and SGD 13

SDG 2: Zero Hunger and SDG 13: Climate Action are two important goals in the 2030 Agenda for Sustainable Development and are directly related to each other (Jain and Mishra 2019).

The nexus between SDG 2: Zero Hunger and SDG 13: Climate Action encompasses multifaceted challenges rooted in their shared commitment to sustainable development. The imperative of sustainable agriculture stands as a crucial convergence, as it not only addresses food security but also combats climate change by curbing greenhouse gas emissions through responsible soil management and efficient farming practices. This interlinkage underscores their collective endeavor to propel sustainable agricultural methods and optimal soil stewardship (Nilsson et al. 2016).

Moreover, climate change wields a direct impact on food security by diminishing crop yields and triggering agricultural losses. Consequently, SDG 2 and SDG 13 jointly advocate for adaptive measures that guarantee food security amidst climate fluctuations. This harmonized pursuit underscores the common goal of preserving access to sustenance amid a changing climate landscape (Dörgő et al. 2018a).

The detrimental ecological implications of food loss and waste further tie these goals together, as their production and transportation processes contribute to greenhouse gas emissions. In this vein, both SDG 2 and SDG 13 spotlight the necessity of curbing food waste while advocating for a more sustainable food ecosystem (Fuso Nerini et al. 2019).

Renewable energy utilization surfaces as a linchpin in SDG 2 and SDG 13's shared vision. By adopting renewable energy sources, greenhouse gas emissions can be mitigated, underscoring these goals' concerted drive towards sustainable agriculture and climate resilience.

Sustainable agriculture, responsible soil management, climate adaptation, curbing food waste, and embracing renewable energy crystallize the binding threads uniting these goals. Through the prism of these concerted efforts, both SDGs navigate the trajectory towards a harmonious coexistence with nature, fostering food security and environmental preservation in tandem (Figure 12).

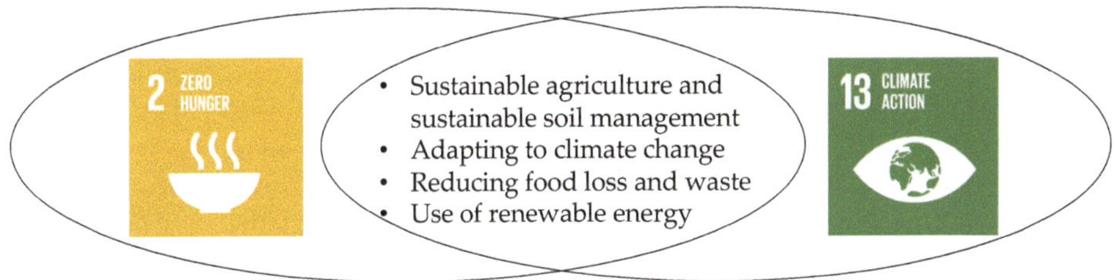

Figure 12. Mutual key elements of SDG 2 and SDG 13. Source: Figure by authors.

Sustainable agriculture can help combat climate change by helping to reduce greenhouse gas emissions through sustainable soil management and the use of more efficient farming methods. Therefore, SDG 2 and SDG 13 focus on promoting sustainable agricultural practices and effective soil management.

Climate change can affect food security, reducing food production and leading to crop losses. Therefore, SDG 2 and SDG 13 have common goals to help communities adapt to and ensure food security in the face of climate change (Dörgő et al. 2018b).

Food loss and food waste contribute to negative impacts on the environment through greenhouse gas emissions generated by production and transport processes. Therefore, SDG 2 and SDG 13 focus on reducing food loss and waste and promoting a more sustainable food system.

Use of renewable energy can help reduce greenhouse gas emissions and combat climate change (Guang-Wen et al. 2023; UN Climate Change 2022). SDG 2 and SDG 13 have common goals to promote the use of renewable energy in agriculture and other sectors.

Thus, it can be concluded that there is a strong and direct link between SDG 2 and SDG 13, which focus on common elements. Sustainable agriculture and sustainable soil management, adapting to climate change, reducing food loss and waste, and using renewable energy are key elements that illustrate this link (Box 12). By promoting sustainable agricultural practices, effective soil management, adapting to climate change, reducing food loss and waste, and using renewable energy, we can contribute to achieving the common goals of the two SDGs and creating a more sustainable food system that ensures food security, reducing the negative impact on the environment (The Lancet 2019).

Box 12. Interlinked actions to reduce hunger and increase climate action.

- *India—Climate change can have a significant impact on India's food security, as the country is heavily dependent on agriculture and vulnerable to extreme changes in temperature and rainfall. At the same time, agriculture in India is responsible for a significant proportion of greenhouse gas emissions. India has, therefore, adopted a number of measures to address these issues, including encouraging organic farming and water-efficient irrigation, promoting healthier and more sustainable diets and developing renewable energy technologies to reduce greenhouse gas emissions from greenhouse gases (Mukhopadhyay et al. 2021).*
- *Brazil—Brazil is the largest producer of soybeans in the world, and the expansion of soybean crops and beef cattle has contributed to the deforestation of the Amazon rainforest. This deforestation has a negative impact on the environment and can lead to lower crop yields and long-term food security issues. At the same time, Brazil is one of the largest emitters of greenhouse gases in the world, and beef production is responsible for a significant proportion of greenhouse gas emissions. Brazil has, therefore, taken steps to address these issues, including imposing stricter restrictions on deforestation and adopting sustainable agricultural practices to reduce greenhouse gas emissions (Stabile et al. 2020).*
- *United States of America—The United States has been affected by extreme weather conditions, such as droughts and violent storms, which can negatively impact food security. At the same time, US agriculture is responsible for a significant proportion of greenhouse gas emissions, and meat production and consumption have been criticized for their environmental impact. To address these issues, the US has begun to promote sustainable agricultural practices as well as the production of healthier and more sustainable foods, such as vegetables and fruits. The US has also begun to adopt measures to reduce greenhouse gas emissions, including switching to renewable energy sources and promoting green transportation (Blackwell and Fellow 2016; United States Department of State and the United States Executive Office of the President, Washington DC 2021).*

3.13. The Link between SDG 2 and SGD 14

SDG 2: Zero Hunger and SDG 14: Life Below Water are two directly interconnected goals that focus on developing a sustainable food system, reducing food insecurity and, for these, protecting marine ecosystems.

The interrelation between SDG 2: Zero Hunger and SDG 14: Life Below Water presents a matrix of challenges intertwined with their shared emphasis on cultivating sustainable food systems, eradicating food insecurity, and safeguarding marine ecosystems. This symbiotic link underscores the paramount importance of shielding and preserving marine food resources, which are vital for the sustenance of coastal communities and the preservation of marine biodiversity. Herein, SDG 2 and SDG 14 harmoniously champion the safeguarding of marine resources through the adoption of sustainable fishing practices and the mitigation of marine pollution, thereby augmenting global food security (Guang-Wen et al. 2023; Maxim and van der Sluijs 2011).

Aquaculture emerges as a pivotal conduit in the journey towards sustainable food production and resource conservation, as it alleviates the strain on marine resources while bolstering food output. In tandem, SDG 2 and SDG 14 champion sustainable aquaculture practices that uphold the tenets of environmental preservation and marine ecosystem protection (Vilalta et al. 2018).

The perilous implications of marine pollution reverberate across food quality and marine habitat integrity, thereby imperiling food security and biodiversity. By targeting reductions in marine pollution through judicious waste and pollution management practices, SDG 2 and SDG 14 jointly advocate for a healthier marine environment (Ward 2006).

Education and heightened public awareness constitute linchpins in fostering transformative shifts in consumer and producer behaviors, thus fostering a sustainable food system and environment (Sullivan et al. 2018; Stephens et al. 2008; Vilalta et al. 2018). This shared ambition to elevate public understanding and consciousness about food security and marine conservation underscores the conjoined objectives of SDG 2 and SDG 14.

Responsible fishing, sustainable aquaculture, pollution reduction, and public education stand as the bedrocks upon which these goals stand united, envisaging a future of sustenance and prosperity firmly rooted in balanced ecological and human stewardship (Figure 13).

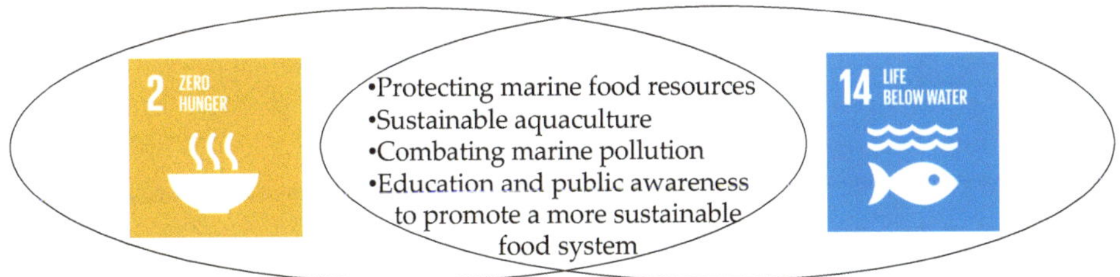

Figure 13. Mutual key elements of SDG 2 and SDG 14. Source: Figure by authors.

Protecting and conserving marine food resources are crucial to ensuring food security for coastal communities and maintaining marine biodiversity. Therefore, SDG 2 and SDG 14 focus on protecting and conserving marine resources by adopting sustainable fishing practices and reducing marine pollution, thereby increasing food security.

Aquaculture is a method of food production that can be used to reduce pressure on marine resources and increase food production (Box 13). Therefore, SDG 2 and SDG 14 have common goals to provide nutritious food, promote sustainable aquaculture practices that respect the principles of environmental conservation, and protect marine ecosystems (Sullivan et al. 2018).

Box 13. Interlinked actions to reduce hunger and increase actions on life below water.

- *Ghana: Ghana is an African country facing food security and marine resource conservation issues. To address these issues, the Ghanaian government has implemented programs to develop sustainable agriculture and aquaculture, promoting ecological agricultural practices and responsible management of marine and coastal resources (FAO 2012; Ministry of Food and Agriculture Republic of Ghana 2018).*
- *Japan: Japan is an island country that depends on marine resources for its food security. The Japanese government has developed policies and programs to conserve marine resources and improve fisheries management, promoting sustainable fishing and implementing measures to protect vulnerable marine areas (Duarte et al. 2020).*
- *Iceland: Iceland is another country that depends on fishing for its food security. At the same time, the Icelandic government has adopted a conservative approach to protect marine resources and conserve fragile marine ecosystems. Iceland has developed and implemented marine resource management policies and programs that promote sustainable fishing and protect vulnerable marine ecosystems (Bryndum-Buchholz et al. 2021).*
- *Brazil: Brazil is a vast country, rich in natural resources, including marine resources. At the same time, Brazil is facing food security problems in some regions of the country. The Brazilian government has developed policies and programs to promote sustainable agriculture and the sustainable management of marine and coastal resources to ensure the sustainable development of the country (Government of Brazil 2022a).*

Marine pollution can affect food quality and lead to the destruction of marine habitats, thereby affecting food security and biodiversity. SDG 2 and SDG 14, therefore, focus on reducing marine pollution and protecting the marine environment by adopting safer waste and pollution management practices.

Education and public awareness are key to promoting a change in behavior among consumers and producers to promote a more sustainable food system and environment. Therefore, SDG 2 and SDG 14 have common objectives to increase the level of education and public awareness in relation to issues of food security and the protection of the marine environment (Stephens et al. 2008).

3.14. The Link between SDG 2 and SGD 15

SDG 2: Zero Hunger and SDG 15: Life on Land are two directly interconnected Sustainable Development Goals that focus on developing a sustainable food system and protecting terrestrial biodiversity.

The interrelation between SDG 2: Zero Hunger and SDG 15: Life on Land unveils a tapestry of challenges interwoven with their shared commitment to forging a sustainable food system and safeguarding terrestrial biodiversity. Soil, a cornerstone of the food system and terrestrial life, stands central within this matrix. SDG 2 and SDG 15 are intrinsically aligned in their pursuit to shield and conserve soil through sustainable agricultural practices while curbing deforestation and activities that foment soil degradation (Maxim and van der Sluijs 2011).

Biodiversity's pivotal role in upholding terrestrial ecosystems and food security underscores the shared purpose of SDG 2 and SDG 15 to preserve biodiversity via sustainable agricultural methods and the safeguarding of natural habitats.

Sustainable agriculture's pivotal importance in realizing SDG 2 and SDG 15 is twofold: alleviating the burden on natural resources and bolstering food production. Hence, SDG 2 and SDG 15 underscore the imperative of sustainable agricultural practices, including the judicious use of pesticides and chemical fertilizers (Maxim and van der Sluijs 2011).

Equally pivotal is the prudent management of natural resources for achieving both SDG 2 and SDG 15. Protection and conservation of resources, such as water, soil, and forests, feature prominently, propelling SDG 2 and SDG 15 to champion improved natural resource management practices that buttress environmental preservation and underpin a sustainable food system.

So, SDG 2 and SDG 15 coalesce to frame a symbiotic interdependence predicated on fashioning a sustainable food system and safeguarding terrestrial biodiversity. The adoptions of sustainable agricultural methodologies, biodiversity protection, astute natural resource stewardship, and soil preservation coalesce as the bedrock of these goals, envisaging a future where the trajectory of humanity is harmonized with the sustenance of both ecosystems and prosperity for all. Consequently, a harmonious collaboration between governments and civil society becomes imperative to concretize these aspirations and usher in an integrated, sustainable era for food security and terrestrial biodiversity (Figure 14).

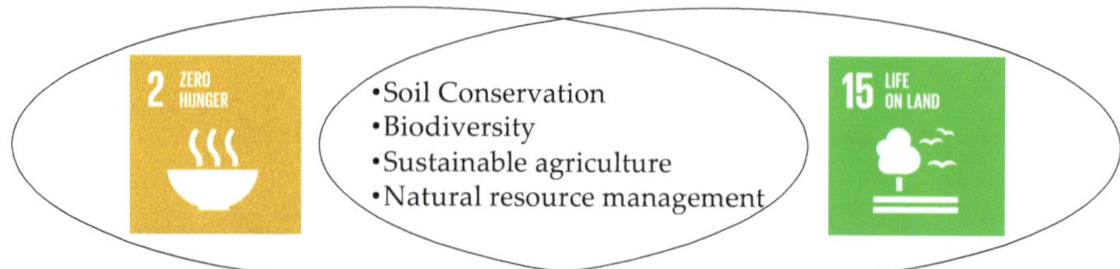

Figure 14. Mutual key elements of SDG 2 and SDG 15. Source: Figure by authors.

Soil is one of the most important elements in the food system and terrestrial biodiversity. Therefore, SDG 2 and SDG 15 focus on protecting and conserving soil through sustainable agricultural practices and avoiding deforestation and other activities that can lead to soil degradation.

Biodiversity is essential for maintaining terrestrial ecosystems and food security. Therefore, SDG 2 and SDG 15 aim to protect biodiversity by adopting sustainable agricultural practices and promoting the conservation of natural habitats (FAO, IFAD, UNICEF, WFP and WHO 2022).

Sustainable agriculture is essential to achieving SDG 2 and SDG 15 by reducing pressure on natural resources and increasing food production. Therefore, SDG 2 and SDG 15 focus on promoting sustainable and sustainable agricultural practices, such as reducing the use of pesticides and chemical fertilizers.

Natural resource management is essential to achieving both SDG 2 and SDG 15, by protecting and conserving natural resources, such as water, soil, and forests. Therefore, SDG 2 and SDG 15 aim to adopt better natural resource management practices to protect the environment and support a sustainable food system (Box **??**).

Box 14. Interlinked actions to reduce hunger and increase actions on life on land.

- *Brazil—Sustainable agriculture and fisheries are important to the Brazilian economy and to the food security of its population. The Brazilian government launched the "ABC Plan" program for agriculture based on sustainable practices and soil and water protection. Brazil has also developed a biodiversity conservation program that aims to protect rainforests and other ecosystems (Ministry of Agriculture, Livestock and Food Supply, Brasília 2021).*
- *Indonesia—Indonesia has some of the greatest biodiversity in the world and an important agricultural economy. The Indonesian government has launched a national food security strategy that promotes sustainable agriculture and the protection of natural resources. In addition, Indonesia launched the "Green Economy" program to promote sustainable economic development and a reduction in greenhouse gas emissions (UN Partnership for Action on Green Economy 2022).*

- *Kenya—Agriculture is an important industry in Kenya, and the Kenyan government has launched the "Agricultural Sector Development Strategy" program to promote sustainable agriculture and food security. In addition, Kenya launched the "Green Economy Strategy and Implementation Plan" program to promote sustainable economic development and biodiversity conservation (African Union Development Agency AUDA-NEPAD 2015).*
- *Norway—Norway is a country with a strong economy based on fishing and aquaculture. The Norwegian government has implemented several measures to protect marine resources and promote sustainable fishing. Norway also has one of the most advanced biodiversity protection policies in the world, including through the creation of protected areas and the promotion of a sustainable way of life (Norway Government 2021; Bjørkhaug and Richards 2008).*

3.15. The Link between SDG 2 and SGD 16

SDG 2: Zero Hunger and SDG 16: Peace, Justice and Strong Institutions are two SDGs that are indirectly interconnected but linked in terms of developing a sustainable food system and ensuring accountable and transparent governance.

The nexus between SDG 2: Zero Hunger and SDG 16: Peace, Justice and Strong Institutions unfolds as a nuanced interplay of challenges and shared aspirations. While indirectly connected, these two goals converge in their dedication to cultivating a sustainable food system and fostering accountable, transparent governance. SDG 2 strives to alleviate food insecurity and ensure universal access to nutritious sustenance, irrespective of social or geographical boundaries. Concurrently, SDG 16 endeavors to nurture social equity, curbing disparities encompassing food accessibility and distribution. Consequently, these goals are interlaced as they collectively pursue the universal provision of wholesome nourishment, irrespective of socioeconomic standing (Dörgő et al. 2018b; Otto-Zimmermann 2012).

SDG 16 pivots on fostering effective governance, transparency, and accountability in public resource management and policy formulation. These tenets are pivotal in shaping sustainable, equitable food policies and practices, thus propelling SDG 2 and SDG 16 in tandem towards a food system underpinned by social justice and upheld by the tenets of transparent governance (Cucurachi and Suh 2017).

Concurrently, conflict and instability can sow seeds of food insecurity by disrupting agricultural endeavors, resource access, and mobility (Ben Hassen and El Bilali 2022). SDG 16's emphasis on peace promotion, conflict prevention, and community stabilization dovetails with SDG 2's mission to alleviate food insecurity. These symbiotic efforts, rooted in addressing underlying inequities and social injustices, stand poised to thwart conflict and instability from germinating.

The linchpin of equitable food resource distribution and a holistic approach to food security is responsible, transparent governance. SDG 2 and SDG 16, thus, rally around bolstering institutional and governmental capabilities to facilitate effective food policy formulation and uphold transparent, accountable governance (Otto-Zimmermann 2012).

In synthesis, SDG 2 and SDG 16, though indirectly intertwined, embolden one another in their quest for sustainable development. Shared aspirations of equitable food access, social justice, effective food governance, conflict prevention, and transparent governance stand as the pillars of their confluence. Crafting a sustainable food system and upholding the tenets of transparent governance emerge as pivotal endeavors in achieving these goals. Thus, the union between SDG 2 and SDG 16 marks a pivotal stride towards constructing a more equitable, sustainable global paradigm (Figure 15).

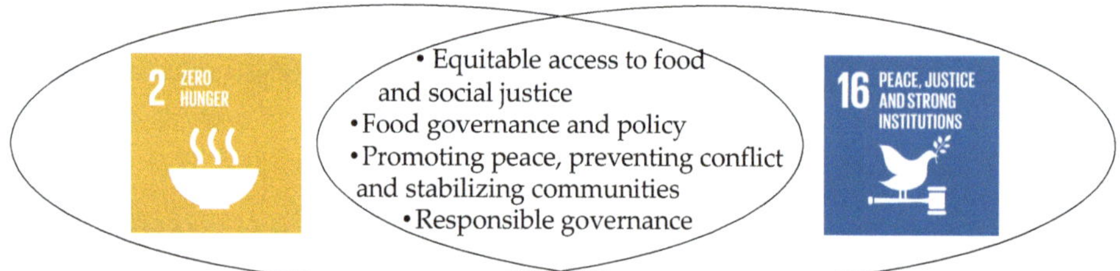

Figure 15. Mutual key elements of SDG 2 and SDG 16. Source: Figure by authors.

SDG 2 aims to reduce food insecurity and increase access to food for all, regardless of social or geographical condition. At the same time, SDG 16 aims to promote social justice and reduce inequalities, including in access to food and its distribution. Thus, the two goals are interconnected in their effort to ensure that everyone has access to sufficient and nutritious food, regardless of their social or economic situation.

SDG 16 focuses on promoting effective governance, transparency, and accountability in the management of public resources and public policies. These efforts are essential to develop sustainable and equitable food policies and practices (FAO, IFAD, UNICEF, WFP and WHO 2022). Therefore, SDG 2 and SDG 16 complement each other in terms of developing a sustainable food system that promotes social justice and respects the principles of good governance.

Conflict and instability can be causes of food insecurity by affecting food production, access to resources, and mobility (Dörgő et al. 2018b; Otto-Zimmermann 2012). Therefore, SDG 16 focuses on promoting peace, preventing conflict, and stabilizing communities, while SDG 2 aims to reduce food insecurity. These efforts complement each other, as conflict and instability can be prevented by addressing underlying inequality and social injustice.

Responsible and transparent governance is essential to ensure the equitable distribution of food resources and an integrated approach to food security (FAO, IFAD, UNICEF, WFP and WHO 2022). SDG 2 and SDG 16, therefore, focus on strengthening institutional and governmental capacities to support the development of effective food policies and ensure accountable and transparent governance.

Therefore, SDG 2 and SDG 16 are two indirectly interconnected sustainable development goals and are complementary. They share a common goal of ensuring a more just, equitable, and sustainable world by promoting access to sufficient and nutritious food for all and by strengthening effective and accountable governance. Equitable access to food and social justice, food governance and policy, peace promotion and conflict prevention, and responsible governance are key elements that illustrate the connection between the two SDGs. Developing a sustainable food system and ensuring transparent and accountable governance are essential to achieving these goals (FAO, IFAD, UNICEF, WFP and WHO 2022). Promoting access to food and strengthening institutional and government capacities are essential efforts in achieving these goals (Box 15).

Box 15. Interlinked actions to reduce hunger and increase levels of peace, justice, and strong institutions.

- *Ghana: In Ghana, the government has implemented agricultural and rural development programs to improve food security and enhance economic growth in rural areas. At the same time, the government has invested in strong institutions and an effective justice system to promote good governance and protect citizens' rights, including the rights to food and water (Asare-Nuamah et al. 2021).*
- *Canada: In Canada, the government has developed programs to support local farmers and promote sustainable food production. These programs helped create a sustainable business environment and protected natural resources, including soil and water. At the same time, Canada has promoted a peaceful and inclusive society through investments in education, health, and other public services, which have helped reduce inequalities and increase social cohesion (Government of Canada 2022).*
- *Rwanda: In Rwanda, the government has implemented policies to improve food production and promote rural development. These policies included investments in rural infrastructure and a better irrigation system, which improved food security and increased agricultural productivity. At the same time, Rwanda invested in strong institutions and promoted good governance to enhance stability and security in the country (Kim et al. 2022).*

3.16. The Link between SDG 2 and SGD 17

SDG 2: Zero Hunger and SDG 17: Partnerships to achieve the Goal are two indirectly linked goals, but they represent one of the most important relations and aim to promote food security and build strong partnerships to achieve the goals.

The interrelation between SDG 2: Zero Hunger and SDG 17: Partnerships to achieve the Goal unveils a paramount association aimed at fostering food security and orchestrating robust collaborations for goal achievement. SDG 17 takes the mantle of mobilizing resources and nurturing alliances to invigorate investments and sustenance for sustainable development, encompassing the fight against food insecurity and malnutrition (Guang-Wen et al. 2023). It also champions innovation in technology and business models to usher in more effective, sustainable food production and distribution. In contrast, SDG 2 is single-minded in its pursuit of curbing food insecurity and heightening access to nourishing sustenance. SDG 17 galvanizes partnerships and cooperation across diverse sectors—the public, private, and civil society—to collectively combat the specters of food insecurity and malnutrition.

Concurrently, SDG 17 underscores the importance of knowledge exchange and propagation of best practices among diverse stakeholders to cultivate sustainable development and an efficacious strategy to address agricultural, food, and nutritional concerns. Thus, SDG 17 and SDG 2, although tangentially linked, synthesize a symbiotic relationship where collaborative endeavors emerge as the bedrock for curtailing food insecurity and malnutrition and propelling the creation of a just, sustainable food system. Mobilizing resources, fostering innovation, intersectoral collaboration, and disseminating knowledge and optimal practices stand as linchpins in this connection, steering the collective journey towards accomplishing sustainable development objectives (Figure 16).

Figure 16. Mutual key elements of SDG 2 and SDG 17. Source: Figure by authors.

SDG 17 aims to mobilize resources and develop strong partnerships to ensure increased investment and support for sustainable development, including to combat food insecurity and malnutrition. Also, it aims to promote innovation in the development of technologies and business models to ensure more efficient and sustainable food production and distribution, while SDG 2 focuses on reducing food insecurity and increasing access to nutritious food. SDG 17 encourages partnerships and collaboration across sectors, including the public sector, the private sector, and civil society, to combat food insecurity and malnutrition (FAO, IFAD, UNICEF, WFP and WHO 2022).

In the meantime, SDG 17 encourages the exchange of knowledge and good practices between different actors to ensure sustainable development and an effective approach to problems related to agricultural production, food, and nutrition (Box 16).

Box 16. Interlinked actions to reduce hunger and enhance partnerships to achieve the goal.

- *Ghana: The Government of Ghana launched the "Planting for Food and Jobs" program in 2017, which aims to increase agricultural production and create jobs for youth and women in rural areas. The program aims to improve productivity, access to markets, and promote a sustainable food system (Asante and Bawakyillenuo 2021).*
- *Brazil: In 2019, Brazil launched the "Casa Verde e Amarela" program, which aims to reduce poverty and inequality in rural areas by investing in infrastructure and increasing access to basic services, such as education, health, and nutrition (Government of Brazil 2022b).*
- *United States: In the United States, the "Feed the Future" program was launched in 2010 to support partner countries in developing sustainable food systems and combating hunger and malnutrition. The program focuses on improving agricultural productivity, access to markets, and capacity to adapt to climate change (The U.S. Government's Global Hunger & Food Security Initiative. Feed the Future n.d.).*
- *India: The Government of India launched the National Food Security Mission in 2007 to improve food security by increasing agricultural production and access to nutritious food for the most vulnerable members of society (Government of India n.d.).*

4. One Goal, Many Results: Discussions on the Role of SDG 2: Zero Hunger in Global Sustainable Development

Since their adoption in 2015, the SDGs have become a unique and ambitious global framework aimed at addressing the complex and interconnected challenges of the modern world. This set of 17 goals and 169 targets represents a global consensus on where the international community should be heading to build a more sustainable and prosperous future for all Earth's inhabitants (UN Transforming Our World: The 2030 Agenda for Sustainable Development n.d.).

The SDGs represent a deep commitment by world leaders to overcome geographic, social, and economic divides and address the common challenges of humanity. These goals reflect the desire to create an environment in which all nations and communities can develop their potential, ensure their well-being, and protect natural resources for future generations (Bell and Morse 2019; Hajer et al. 2015; Malešević Perović and Mihaljević Kosor 2020). Thus, the SDGs represent a global vision of a fairer and more inclusive future, where every individual has access to a dignified life and equal opportunities.

In a world characterized by complex and interconnected challenges, such as climate change, social and economic inequalities, political instability, and global health issues, the SDGs have become the anchor to which the efforts of nations and international organizations to address these challenges are reported. They provide a platform for collaboration, innovation, and joint action, with the goal of creating sustainable and scalable solutions (Bell and Morse 2019; Swain 2018; Hajer et al. 2015).

A key aspect of the SDGs is their holistic approach and the interconnectedness of the goals. Each objective does not act in isolation but influences and is influenced by the others (Fonseca et al. 2020; Nilsson et al. 2016). This approach recognizes that global challenges are often interrelated and that effective solutions require an integrated perspective. For example, goals such as health and education are closely related to food security and the sustainable management of natural resources (Malešević Perović and Mihaljević Kosor 2020). This interconnectedness requires cooperation and coordination between different sectors and levels of government to achieve meaningful results.

One of the great lessons learned from the adoption of the SDGs is that the world's challenges cannot be tackled in isolation. They intersect and intertwine, and solutions for one can influence the achievement of the others. SDG 2: Zero Hunger is a perfect illustration of this interconnectedness, impacting and, in turn, being influenced by all the other SDGs (Cheo and Kugedera 2021).

For example, the link between SDG 2: Zero Hunger and SDG 1: Eradication of Poverty is evident in the fight against food poverty. Poverty is a determinant of food insecurity and achieving SDG 2: Zero Hunger is essential to improving the health and well-being of populations affected by poverty.

In the same way, the connection between SDG 2: Zero Hunger and SDG 3: Health and Well-being is profound. Access to healthy and nutritious food has a significant impact on population health. Conversely, poor health can affect a person's ability to provide adequate nutrition, creating a vicious cycle of vulnerability.

Education and training are key elements in achieving the SDGs, and SDG 2: Zero Hunger and SDG 4: Quality Education are interconnected by promoting awareness of nutrition and sustainable agriculture. Education improves knowledge of efficient agricultural practices, stimulating innovation and more sustainable approaches.

SDG 2: Zero Hunger also influences SDG 5: Sanitation and Clean Water. Access to potable water and proper hygiene are essential to ensuring food security. Improper sanitation can lead to food contamination and health problems, which underscores the need for an integrated approach to these goals.

On the other hand, SDG 2: Zero Hunger and SDG 13: Combating Climate Change influence each other. Climate change can affect food production and the availability of agricultural resources, putting food security at risk. At the same time, sustainable agricultural practices can contribute to reducing greenhouse gas emissions and adapting to climate change (The Lancet 2019).

Achieving the SDGs is a complex challenge, and SDG 2: Zero Hunger plays a crucial role in this equation. The interconnections between SDG 2: Zero Hunger and the other goals illustrate that addressing global challenges requires a holistic and collaborative approach. The transformation targeted by the SDGs is deeply interdependent, where each goal influences and is influenced by the others. By understanding these complex connections, the global community can move towards a more sustainable and equitable future for all.

The SDGs represent an important milestone in humanity's efforts to embark on a sustainable and responsible path. They reflect a shared vision for a better future and a commitment to act in a coherent and concerted way to achieve these goals. The link between SDG 2: Zero Hunger and the other goals illustrates that every step forward in achieving a goal brings significant benefits for the entire planet and for future generations (Griggs et al. 2013). The implementation of the SDGs requires political will, global collaboration, and individual commitment to turn the vision into reality (Bakshi et al. 2018; Bell and Morse 2019).

We are currently at a pivotal moment in global efforts to achieve the SDGs as a firm commitment to build a fairer and more sustainable future for everyone. While significant progress has been made in many regions and areas, it is clear that action to achieve the SDGs is not yet advancing at the speed or scale needed to address the complex and interconnected challenges facing humanity.

One of the central objectives of the SDGs is ending hunger and food insecurity (SDG 2: Zero Hunger). This challenge is critical in a world where approximately

9% of the population suffers from chronic undernourishment, and risks associated with climate change, conflict, and economic inequality threaten global food security. Although there have been improvements in access to food and undernutrition, progress is still insufficient to achieve the goal of ending hunger by 2030 (UN Progress towards the Sustainable Development Goals n.d.; The Lancet 2019).

In many regions of the world, government programs and innovations in agriculture have helped increase food production and improve food security. However, persistent inequalities and poverty remain major obstacles to achieving SDG 2: Zero Hunger. Also, climate change and the risks associated with extreme events disproportionately affect vulnerable communities, putting food security at risk.

The SDGs are a global effort to address the complex and interconnected challenges of the modern world. While progress is being made in many areas, there is still an urgent need to accelerate action and address inequalities and vulnerabilities (UN Progress towards the Sustainable Development Goals n.d.). SDG 2: Zero Hunger is emblematic of the challenges and opportunities that the SDGs bring as a whole. Understanding the links between these goals and promoting coordinated and coherent action are essential to building a better and more sustainable future for all.

The importance of the holistic link between SDG 2 Zero Hunger and all other SDGs lies in the recognition that sustainable development cannot be achieved in isolation but requires an integrated and coordinated approach to the complex and interconnected issues facing contemporary society (Bakshi et al. 2018). Understanding the synergies between SDG 2 and the other SDGs is essential to maximizing the impact of our actions and creating a fairer and more prosperous future for everyone (Figure 17).

By interlinking SDG 2: Zero Hunger with the other SDGs, it is recognized that hunger and food insecurity cannot be eliminated in isolation from other global challenges, such as climate change, poverty, inequalities, and environmental degradation (Guang-Wen et al. 2023; Dörgő et al. 2018a). Approaching SDG 2: Zero Hunger within the broader context of sustainable development allows us to explore the ways in which reducing hunger can contribute to solving complex problems, and vice versa (The Lancet 2019).

This link is also crucial to avoid the unwanted negative effects of measures taken within one objective on other areas. For example, increasing food production to achieve the goal of SDG 2: Zero Hunger must be balanced with the need to conserve natural resources and the environment, according to SDG 15: Life on Land and SDG 14: Life Below Water (Salasan and Balan 2022).

Figure 17. Links between SDG 2: Zero Hunger and all other SDGs. Source: Figure by authors.

The link between SDG 2: Zero Hunger and the other goals is a recognition that food security and equity are crucial to promoting human well-being, and these issues are closely connected to health (SDG 3), education (SDG 4), and economic growth (SDG 8). Promoting a sustainable and equitable food system has significant implications on poverty, inequalities, and the general well-being of the population, according to SDG 1 (No Poverty) and SDG 10 (Reduced Inequalities) (Bastianoni et al. 2018).

Therefore, the importance of the holistic link between SDG 2: Zero Hunger and the other SDGs is essential for genuine and effective sustainable development. This approach allows us to see the subtle and complex connections between different global challenges and identify solutions that have a positive impact in multiple areas (Guang-Wen et al. 2023).

Despite progress in some areas, actions to achieve the SDGs are not progressing uniformly or in sync across all sectors and regions. This can lead to imbalances and undermine global efforts (Bastianoni et al. 2018). For example, progress in education (SDG 4) can have a positive impact on health (SDG 3), and sustainable economic growth (SDG 8) can help reduce inequalities (SDG 10) (UN Progress towards the Sustainable Development Goals n.d.; Inter-Agency and Expert Group on Sustainable Development Goal Indicators (IAEG-SDGs) n.d.).

The reasons for the slow pace of progress are varied and complex. One aspect is the lack of adequate funding and resources to implement specific measures (Spangenberg 2017; Maxim and van der Sluijs 2011). Also, the need for global and national coordination to ensure synergies between different agencies and sectors is essential (Guang-Wen et al. 2023). In addition, political instability, conflicts, and crisis situations can make it difficult to implement the SDGs in certain regions (Ben Hassen and El Bilali 2022).

However, it is important to emphasize that we are not at an impasse. Despite the challenges, there have been significant successes in areas, such as health, education, and poverty reduction (World Health Organization 2016).

Drawing insights from experiences and adapting strategies, the acceleration of the SDG achievement becomes attainable. The promotion of innovation, community engagement, and international collaboration holds the potential to expedite progress and facilitate the establishment of a sustainable future (Fuso Nerini et al. 2018). Through these means, the groundwork is laid for the cultivation of a future characterized by equity, prosperity, and sustainability across global communities and ecosystems.

5. Holistic Solutions: Concluding Reflections on SDG 2: Zero Hunger and Its Links with the SDGs

The challenges linked to SDG 2: Zero Hunger resonate across the entire spectrum of SDGs, either through direct connections or more nuanced and indirect correlations, necessitating a comprehensive and intricate approach for their comprehension. The centrality of SDG 2: Zero Hunger to the accomplishment of other objectives is evident in its pivotal role in safeguarding people's health and well-being, inherently tied to the availability of adequate and nourishing sustenance. Effectively tackling the problem of hunger and malnutrition demands a comprehensive strategy addressing underlying concerns, such as poverty, inequality, climate change, and sustainable economic growth, all of which hold intricate ties with each of the remaining SDGs (UN Progress towards the Sustainable Development Goals n.d.).

Addressing poverty and inequality can lead to a decrease in hunger by expanding access to nourishing food sources. Concurrently, countering climate change supports sustainable agricultural and ecosystem development, thus assuring the stability of food supplies and water resources (WHO/UNICEF 2008). Similarly, educational initiatives and sustainable economic growth foster employment opportunities, particularly in sectors like agriculture and food production, bolstering food security. The fight against food wastage and losses contributes not only to SDG 12's aim of sustainable consumption and production but also to curbing environmental impacts through reduced greenhouse gas emissions and judicious resource utilization (Francis and McDonagh 2016).

Consequently, attaining the aspirations of SDG 2: Zero Hunger requires an integrated and multifaceted approach, encompassing the complexities of poverty alleviation, inequality reduction, climate change mitigation, sustainable economic development, and waste reduction in tandem with fostering sustainable consumption and production (Le Blanc 2015).

To achieve the aims of SDG 2: Zero Hunger, a widespread and synergistic engagement of global, regional, and local authorities becomes paramount. The objective cannot be fulfilled through isolated endeavors but necessitates synchronized cooperation between governmental bodies, international organizations, civil society, private enterprises, and other stakeholders (Le Blanc 2015; Griggs et al. 2013).

At the global stage, governments and international organizations must formulate policies and programs addressing the root causes of hunger while advocating for sustainable agriculture and balanced diets. These initiatives should be seamlessly integrated into national developmental agendas, subjected to regular monitoring and evaluation to gauge their effectiveness (Swain 2018). Regionally and nationally, comprehensive strategies should be devised to curtail poverty, bolster educational and healthcare accessibility, and endorse sustainable agricultural practices and healthy nutrition. In this context, public–private partnerships should be established to ensure effective strategy implementation. Locally, authorities must frame policies aligning with the specific needs of communities, endorsing sustainable agriculture and nutritional practices. Collaborations between governing bodies, local communities, and private entities are essential to yield context-specific solutions (Griggs et al. 2013; Reyers et al. 2017).

The staggering fact that all the food produced but left uneaten could feasibly feed around two billion people, and this exceeds double the count of undernourished people worldwide, acutely underscores not just a systemic failure but a disconcerting lack of prioritization when it comes to combating hunger (World Food Programme 2020). While SDG 2: Zero Hunger ambitiously aims to achieve zero hunger, the stark reality remains that despite these aspirations, the measures put in place to mitigate food waste often fall far short of their intended impact. The inefficiency of food waste reduction strategies not only speaks to a lack of resourcefulness and effective implementation but also exposes a disquieting apathy towards the prevailing global issue of food insecurity (Ivanova et al. 2016; Balan et al. 2022b; Feher et al. 2021). Indeed, food waste can be perceived as a manifestation of indifference and, in some aspects, even a form of selfishness. It is a stark reminder of how excess and disregard can coexist within the same geographical boundaries. Contrary to a common assumption, geographical distance is not the sole culprit; food waste and hunger are present simultaneously, even within the confines of the same country, city, or even neighborhood. This poignant reality highlights a disheartening paradox where food waste occurs side by side with food insecurity, illustrating a stark disparity between those who are wasteful and those who suffer from hunger. In such instances, food wastage is not just an act of negligence; it is a poignant demonstration of disregard for the plight of those in need, an inadvertent endorsement of an alarming culture of neglect and indifference to the global food crisis. People-to-people collaborations in local communities are essential to achieve context-specific solutions. Therefore, the education of the population in communities of all sizes, regarding food waste, is deficient, as long as everyone has neighbors or knows people who suffer from hunger while food is wasted in their own homes.

The realization of SDG 2: Zero Hunger hinges on involving stakeholders at every level and harmonizing efforts globally, regionally, nationally, and locally (FAO, IFAD, UNICEF, WFP and WHO 2023). A holistic approach is imperative to address hunger's root causes and propagate sustainable agriculture and nutritious dietary habits.

Simultaneously, the continuous assessment of advancements is essential to gauge the extent of SDG 2: Zero Hunger's accomplishments. Monitoring aids in identifying vulnerabilities and shortcomings, contributing to the development of effective policies and interventions against hunger and malnutrition (World Food Programme. Hunger Map 2022). Assessment should span topics, such as food accessibility, equal food distribution, nutritional standards, health, and education. Up-to-date relevant data are pivotal for informed decision making and tracking progress over time. Moreover, monitoring and evaluation must be conducted across local, national, and international tiers for a comprehensive grasp of the situation. Communication and cooperation among stakeholders at various levels remain crucial for sharing information and refining decision-making processes (FAO, IFAD, UNICEF, WFP and WHO 2023).

Conversely, the shortcomings and limitations of SDG 2: Zero Hunger come to light in its failure to encapsulate the intricacies of food systems and the food industry, both instrumental in either exacerbating or mitigating hunger and malnutrition. Additionally, the SDG does not explicitly address environmental concerns or the implications of climate change on food security. Some of its objectives and metrics lack specificity or fall short of offering comprehensive measurement tools (Halkos and Gkampoura 2021; Sachs et al. 2019).

Given these circumstances, it is undeniable that monitoring and assessing the progress of SDG 2: Zero Hunger, its interlinkages with other SDGs, and their reciprocating impact are pivotal in refining the current scenario and formulating effective policies and programs in the battle against hunger and malnutrition.

Author Contributions: Conceptualization, R.L. and I.M.B.; Methodology, M.O.; Software, M.O.; Validation, R.L., I.M.B. and M.O.; Formal Analysis, M.O.; Investigation, I.M.B.; Resources, R.L.; Data Curation, M.O.; Writing—Original Draft Preparation, R.L.; Writing—Review and Editing, I.M.B.; Visualization, M.O; Supervision, R.L.; Project Administration, R.L.; Funding Acquisition, M.O.

Funding: This research received no external funding.

Conflicts of Interest: The authors declare no conflict of interest.

References

African Union Development Agency AUDA-NEPAD. 2015. Kenya Signs Continental Agricultural Improvement Scheme. Available online: https://nepad.org/news/kenya-signs-continental-agricultural-improvement-scheme#:~:text=Kenya%2520has%2520officially%2520launched%2520its,domesticating%2520the%2520continental%2520agricultural%2520program (accessed on 20 March 2023).

Ali, Umar Mohammed. 2022. Demerits side of open defecation on school education in nigeria. *Frontline Social Sciences and History Journal* 2: 28–33. [CrossRef]

Anser, Muhammad Khalid, Muhammad Azhar Khan, Abdelmohsen A. Nassani, Abdullah Mohammed Aldakhil, Xuan Hinh Voo, and Khalid Zaman. 2021. Relationship of environment with technological innovation, carbon pricing, renewable energy, and global food production. *Economics of Innovation and New Technology* 30: 807–42. [CrossRef]

Arshad, Ashraf, and Jamil Khalid. 2022. Solar-powered irrigation system as a nature-based solution for sustaining agricultural water management in the Upper Indus Basin. *Nature-Based Solutions* 2: 100026. [CrossRef]

Asante, Felix Ankomah, and Simon Bawakyillenuo. 2021. *Farm-Level Effects of the 2019 Ghana Planting for Food and Jobs Program: An Analysis of Household Survey Data*. GSSP Working Paper 57. Washington, DC: International Food Policy Research Institute (IFPRI). [CrossRef]

Asare-Nuamah, Peter, Anthony Amoah, and Simplice Asongu. 2021. Achieving Food Security in Ghana: Does Governance Matter? *Politics & Policy* 51: 614–35. [CrossRef]

Australian Government. National Food Waste Strategy. 2017. Halving Australia's Food Waste by 2030. Available online: https://www.agriculture.gov.au/sites/default/files/documents/national-food-waste-strategy.pdf (accessed on 1 May 2023).

Bakshi, Bhavik R., Timothy G. Gutowski, and Dusan P. Sekulic. 2018. Claiming Sustainability: Requirements and Challenges. *ACS Sustainable Chemistry & Engineering* 6: 3632–39. Available online: https://pubs.acs.org/doi/abs/10.1021/acssuschemeng.7b03953 (accessed on 13 February 2023).

Balan, Ioana Mihaela, Emanuela Diana Gherman, Ioan Brad, Remus Gherman, Adina Horablaga, and Teodor Ioan Trasca. 2022a. Metabolic Food Waste as Food Insecurity Factor—Causes and Preventions. *Foods* 11: 2179. [CrossRef] [PubMed]

Balan, Ioana Mihaela, Emanuela Diana Gherman, Remus Gherman, Ioan Brad, Raul Pascalau, Gabriela Popescu, and Teodor Ioan Trasca. 2022b. Sustainable Nutrition for Increased Food Security Related to Romanian Consumers' Behavior. *Nutrients* 14: 4892. [CrossRef] [PubMed]

Bastianoni, Simone, Luca Coscieme, Dario Caro, Nadia Marchettini, and Federico M. Pulselli. 2018. The needs of sustainability: The overarching contribution of systems approach. *Ecological Indicators. in press.* [CrossRef]

Bell, Simon, and Stephen Morse. 2019. Sustainability Indicators Past and Present: What Next? *Sustainability* 10: 1688. [CrossRef]

Ben Hassen, Tarek, and Hamid El Bilali. 2022. Impacts of the Russia-Ukraine War on Global Food Security: Towards More Sustainable and Resilient Food Systems? *Foods* 11: 2301. [CrossRef]

Bielski, Stanisław, Anna Zielińska-Chmielewska, and Renata Marks-Bielska. 2021. Use of Environmental Management Systems and Renewable Energy Sources in Selected Food Processing Enterprises in Poland. *Energies* 14: 3212. [CrossRef]

Bjørkhaug, Hilde, and Carol Ann Richards. 2008. Multifunctional agriculture in policy and practice? A comparative analysis of Norway and Australia. *Journal of Rural Studies* 24: 98–111. [CrossRef]

Blackwell, Ashley, and Emerson Fellow. 2016. Center for American Progress. Best Practices for Creating a Sustainable and Equitable Food System in the United States. Available online: https://www.americanprogress.org/article/best-practices-for-creating-a-sustainable-and-equitable-food-system-in-the-united-states/ (accessed on 10 March 2023).

Bryndum-Buchholz, Andrea, Derek P. Tittensor, and Heike K. Lotze. 2021. The status of climate change adaptation in fisheries management: Policy, legislation and implementation. *Fish and Fisheries* 22: 1248–73. [CrossRef]

Chen, Chia-Yang, Sheng-Wei Wang, Hyunook Kim, Shu-Yuan Pan, Chihhao Fan, and Yupo J. Lin. 2021. Non-conventional water reuse in agriculture: A circular water economy. *Water Research* 199: 117193. [CrossRef] [PubMed]

Cheo, Ambe Emmanuel, and Andrew Tapiwa Kugedera. 2021. Understanding the Interlinkages between SDG-2 and the Other SDGs. In *SDG 2—Zero Hunger: Food Security, Improved Nutrition and Sustainable Agriculture (Concise Guides to the United Nations Sustainable Development Goals)*. Bingley: Emerald Publishing Limited, pp. 43–48. [CrossRef]

Cohen, Matthew. 2017. A Systematic Review of Urban Sustainability Assessment Literature. *Sustainability* 9: 2048. [CrossRef]

Cucurachi, Stefano, and Sangwon Suh. 2017. Cause-effect analysis for sustainable development policy. *Environmental Reviews* 25: 358–79. [CrossRef]

Da Silva, José Graziano, Mauro Eduardo Del Grossi, and Caio Galvão de França. 2011. *The Fome Zero (Zero Hunger) Program: The Brazilian Experience*. Brasília: Ministry of Agrarian Development. Available online: https://www.fao.org/3/i3023e/i3023e.pdf (accessed on 10 March 2023).

De Souza, Sthephany Rayanne Gomes, Diôgo Vale, Hortência Ingreddys Fernandes do Nascimento, Juliano Capelo Nagy, Antônio Hermes Marques da Silva Junior, Priscilla Moura Rolim, and Larissa Mont'Alverne Jucá Seabra. 2023. Food Purchase from Family Farming in Public Institutions in the Northeast of Brazil: A Tool to Reach Sustainable Development Goals. *Sustainability* 15: 2220. [CrossRef]

Di Fabio, Annamaria, and Marc A. Rosen. 2020. An Exploratory Study of a New Psychological Instrument for Evaluating Sustainability: The Sustainable Development Goals Psychological Inventory. *Sustainability* 12: 7617. [CrossRef]

DigiFarm: A Digital Platform for Farmers. n.d. Available online: https://mercycorpsagrifin.org/wp-content/uploads/2019/05/DigiFarm-Platform-Case_Final_pdf (accessed on 15 March 2023).

Dora, Carlos, Andy Haines, John Balbus, Elaine Fletcher, Heather Adair-Rohani, Graham Alabaster, Rifat Hossain, Mercedes De Onis, Francesco Branca, and Maria Neira. 2015. Indicators linking health and sustainability in the post-2015 development agenda. *The Lancet* 385: 380–91. [CrossRef]

Dörgő, Gyula, Gergely Honti, and János Abonyi. 2018a. Automated analysis of the interactions between sustainable development goals extracted from models and texts of sustainability science. *Chemical Engineering Transactions* 70: 781–86. [CrossRef]

Dörgő, Gyula, Viktor Sebestyén, and János Abonyi. 2018b. Evaluating the Interconnectedness of the Sustainable Development Goals Based on the Causality Analysis of Sustainability Indicators. *Sustainability* 10: 3766. [CrossRef]

Duarte, Carlos M., Susana Agusti, Edward Barbier, Gregory L. Britten, Juan Carlos Castilla, Jean-Pierre Gattuso, Robinson W. Fulweiler, Terry P. Hughes, Nancy Knowlton, Catherine E. Lovelock, and et al. 2020. Rebuilding marine life. *Nature* 580: 39–51. [CrossRef] [PubMed]

Egbuikwem, Precious N., Jose C. Mierzwa, and Devendra P. Saroj. 2020. Assessment of suspended growth biological process for treatment and reuse of mixed wastewater for irrigation of edible crops under hydroponic conditions. *Agricultural Water Management* 231: 106034. [CrossRef]

FAO. 2012. Ghana National Aquaculture Development Plan (GNADP). Available online: https://faolex.fao.org/docs/pdf/gha149443.pdf (accessed on 15 March 2023).

FAO. 2015. Scaling Up The Brazilian School Feeding Model. Available online: https://www.fao.org/3/i4287e/i4287e.pdf (accessed on 15 March 2023).

FAO, IFAD, UNICEF, WFP and WHO. 2020. *The State of Food Security and Nutrition in the World 2020: Transforming Food Systems for Affordable Healthy Diets*. Rome: FAO. Available online: https://www.fao.org/3/ca9692en/ca9692en.pdf (accessed on 10 March 2023).

FAO, IFAD, UNICEF, WFP and WHO. 2021. *The State of Food Security and Nutrition in the World 2021: Transforming Food Systems for Food Security, Improved Nutrition and Affordable Healthy Diets for All*. Rome: FAO. Available online: https://www.fao.org/3/cb4474en/cb4474en.pdf (accessed on 10 March 2023).

FAO, IFAD, UNICEF, WFP and WHO. 2022. *The State of Food Security and Nutrition in the World 2022: Repurposing Food and Agricultural Policies to Make Healthy Diets More Affordable*. Rome: FAO. Available online: https://www.fao.org/3/cc0639en/cc0639en.pdf (accessed on 15 March 2023).

FAO, IFAD, UNICEF, WFP and WHO. 2023. *The State of Food Security and Nutrition in the World 2023: Urbanization, Agrifood Systems Transformation and Healthy Diets across the Rural–Urban Continuum*. Rome: FAO, ISSN 2663-807X. Available online: https://www.fao.org/3/cc3017en/cc3017en.pdf (accessed on 10 February 2023).

Farm-to-School USA. n.d. Delaware State. Available online: https://agriculture.delaware.gov/communications-marketing/farm-to-school/ (accessed on 15 March 2023).

Feher, Andrea, Tiberiu Iancu, Mroslav Raicov, and Adrian Banes. 2021. Food Consumption and Ways to Ensure Food Security in Romania. In *Shifting Patterns of Agricultural Trade*. Singapore: Springer. Available online: https://link.springer.com/chapter/10.1007/978-981-16-3260-0_13 (accessed on 10 March 2023).

Fonseca, Luis Miguel, José Pedro Domingues, and Alina Mihaela Dima. 2020. Mapping the Sustainable Development Goals Relationships. *Sustainability* 12: 3359. [CrossRef]

Fosu, Augustin Kwasi. 2015. Growth, inequality and poverty in Sub-Saharan Africa: Recent progress in a global context. *Oxford Development Studies* 43: 44–59. [CrossRef]

Francis, Pope, and Sean McDonagh. 2016. *On Care for Our Common Home: Laudato Si'—The Encyclical of Pope Francis on the Environment*. New York: Publisher Orbis Books (Ecology and Justice), ISBN-10: 1626981736, ISBN-13:978-1626981737.

Fuso Nerini, Francesco, Benjamin Sovacool, Nick Hughes, Laura Cozzi, Ellie Cosgrave, Mark Howells, Massimo Tavoni, Julia Tomei, Hisham Zerriffi, and Ben Milligan. 2019. Connecting climate action with other Sustainable Development Goals. *Nature Sustainability* 2: 674–680. [CrossRef]

Fuso Nerini, Francesco, Julia Tomei, Long Seng To, Iwona Bisaga, Priti Parikh, Mairi Black, Aiduan Borrion, Catalina Spataru, Vanesa Castán Broto, Gabrial Anandarajah, and et al. 2018. Mapping synergies and trade-offs between energy and the Sustainable Development Goals. *Nature Energy* 3: 10–15. [CrossRef]

George, Neetu Abey, and Fiona H. McKay. 2019. The public distribution system and food security in India. *International Journal of Environmental Research and Public Health* 16: 3221. [CrossRef]

Ghana School Feeding Programme (GSFP). n.d. Available online: https://www.mogcsp.gov.gh/ghana-school-feeding-programme-gsfp/ (accessed on 15 April 2023).

Good Food Purchasing Program USA. n.d. Available online: https://goodfoodpurchasing.org (accessed on 15 April 2023).

Government of Brazil. 2022a. Agenda for a More Sustainable Brazil. ISBN 978-65-86360-52-3. Available online: https://www.google.com/url?sa=i&rct=j&q=&esrc=s&source=web&cd=&ved=0CDgQw7AJahcKEwig_fn2_fyAAxUAAAAAHQAAAAAQAw&url=https%253A%252F%252Fwww.gov.br%252Fpt-br%252Fcampanhas%252Fbrasil-na-cop%252Fcopy2_of_absengwebv2.pdf%252F%2540%2540download%252Ffile&psig=AOvVaw31vEG3jGHV52GFMn0vV80N&ust=1693230670697287&opi=89978449 (accessed on 15 April 2023).

Government of Brazil. 2022b. Housing for Vulnerable Communities: Casa Verde e Amarela. Available online: https://www.gov.br/en/government-of-brazil/latest-news/2022/casa-verde-e-amarela (accessed on 1 April 2023).

Government of Canada. 2022. The Government of Canada Invests in Clean Technology to Support Sustainable Farming Practices. Available online: https://www.canada.ca/en/agriculture-agri-food/news/2022/02/the-government-of-canada-invests-in-clean-technology-to-support-sustainable-farming-practices0.html (accessed on 13 March 2023).

Government of India. n.d. Ministry of Agriculture & Farmers Welfare. National Food Security Mission. Available online: https://www.nfsm.gov.in (accessed on 15 April 2023).

Government of the Netherlands. Circular Dutch Economy by 2050. n.d. Available online: https://www.government.nl/topics/circular-economy/circular-dutch-economy-by-2050#:~:text=The%2520Netherlands%2520aims%2520to%2520have,and%2520raw%2520materials%2520are%2520reused (accessed on 1 April 2023).

Griggs, David, Mark Stafford-Smith, O. Gaffney, Johan Rockström, Marcus C. Öhman, Pryia Shyamsundar, Will Steffen, Gisbert Glaser, Norichika Kanie, and Ian Noble. 2013. Policy: Sustainable development goals for people and planet. *Nature* 495: 305–7. [CrossRef]

Guachalla, Francisco J. A. 2023. Embracing the SDG 2030 and Resilience for Monitoring and Learning in Emergency and Developing Projects. *Medical Sciences Forum* 19: 6. [CrossRef]
Guang-Wen, Zheng, Muntasir Murshed, Abu Bakkar Siddik, Md Shabbir Alam, Daniel Balsalobre-Lorente, and Haider Mahmood. 2023. Achieving the objectives of the 2030 sustainable development goals agenda: Causalities between economic growth, environmental sustainability, financial development, and renewable energy consumption. *Sustainable Development* 31: 680–97. [CrossRef]
Hajer, Maarten, Måns Nilsson, Kate Raworth, Peter Bakker, Frans Berkhout, Yvo De Boer, Johan Rockström, Kathrin Ludwig, and Marcel Kok. 2015. Beyond Cockpit-ism: Four Insights to Enhance the Transformative Potential of the Sustainable Development Goals. *Sustainability* 7: 1651–60. [CrossRef]
Halkos, George, and Eleni-Christina Gkampoura. 2021. Where do we stand on the 17 Sustainable Development Goals? An overview on progress. *Economic Analysis and Policy* 70: 94–122. [CrossRef]
Hoddinott, John, and Lawrence Haddad. 1995. Does female income share influence household expenditures? Evidence from Côte d'Ivoire. *Oxford Bulletin of Economics and Statistics* 57: 77–96. [CrossRef]
Hutton, Guy, and Laurence Haller. 2004. *Sanitation Water, and World Health Organization. Evaluation of the Costs and Benefits of Water and Sanitation Improvements at the Global Level.* No. WHO/SDE/WSH/04.04. Geneva: World Health Organization. Available online: https://scholar.google.com/scholar?hl=en&as_sdt=0%252C5&q=Hutton%252C+Guy%252C+Laurence+Haller.+2004.+Sanitation+Water%252C+and+World+Health+Organization.+Evaluation+of+the+costs+and+benefits+of+water+and+sanitation+improvements+at+the+global+level.+No.+WHO%252FSDE%252FWSH%252F04.04.+World+Health+Organization&btnG= (accessed on 15 April 2023).
Index of Sustainable Development of Cities—Brazil. n.d. Brasil IDSC—BR. Available online: https://idsc.cidadessustentaveis.org.br/# (accessed on 15 April 2023).
Inter-Agency and Expert Group on Sustainable Development Goal Indicators (IAEG-SDGs). n.d. Global Indicator Framework for the Sustainable Development Goals and Targets of the 2030 Agenda for Sustainable Development. Available online: https://unstats.un.org/sdgs/indicators/Global%2520Indicator%2520Framework%2520after%25202023%2520refinement_Eng.pdf (accessed on 1 July 2023).
International Fund for Agricultural Development—Bangladesh. n.d. Available online: https://www.ifad.org/en/web/operations/w/country/bangladesh (accessed on 10 March 2023).
Ivanova, Diana, Konstantin Stadler, Kjartan Steen-Olsen, Richard Wood, Gibran Vita, Arnold Tukker, and Edgar G. Hertwich. 2016. Environmental impact assessment of household consumption. *Journal of Industrial Ecology* 20: 526–36. [CrossRef]

Jain, Ashok Kumar, and Sundar Narayan Mishra. 2019. Role of NITI Aayog in the Implementation of the 2030 Agenda. In *2030 Agenda and India: Moving from Quantity to Quality*. Singapore: Springer. [CrossRef]

Janoušková, Svatava, Tomáš Hák, and Bedřich Moldan. 2018. Global SDGs Assessments: Helping or Confusing Indicators? *Sustainability* 10: 1540. [CrossRef]

Jouzdani, Javid, and Kannan Govindan. 2021. On the sustainable perishable food supply chain network design: A dairy products case to achieve sustainable development goals. *Journal of Cleaner Production* 278: 123060. [CrossRef]

Kayser, Georgia L., Namratha Rao, Rupa Jose, and Anita Raj. 2019. Water, sanitation and hygiene: Measuring gender equality and empowerment. *Bulletin of the World Health Organization* 97: 438. [CrossRef]

Kilimo Biashara Program, Kenya. n.d. Available online: https://kilimotrust.org (accessed on 15 April 2023).

Kim, Sung Kyu, Fiona Marshall, and Neil M. Dawson. 2022. Revisiting Rwanda's agricultural intensification policy: Benefits of embracing farmer heterogeneity and crop-livestock integration strategies. *Food Security* 14: 637–56. [CrossRef]

Le Blanc, David. 2015. Towards integration at last? The sustainable development goals as a network of targets. *Sustainable Development* 23: 176–87. [CrossRef]

Lucato, Wagner Cezar, José Carlos da Silva Santos, and Athos Paulo Tadeu Pacchini. 2018. Measuring the Sustainability of a Manufacturing Process: A Conceptual Framework. *Sustainability* 10: 81. [CrossRef]

Maternal, Child Health and Nutrition Project, Nepal. n.d. Available online: https://adranepal.org/impact/health/maternal-health-and-child-nutrition-project/ (accessed on 13 March 2023).

Maxim, Laura, and Jeroen P. van der Sluijs. 2011. Quality in environmental science for policy: Assessing uncertainty as a component of policy analysis. *Environmental Science & Policy* 14: 482–92. [CrossRef]

Mbatha, Mfaniseni Wiseman, Hlanganani Mnguni, and Mandla A. Mubecua. 2021. Subsistence farming as a sustainable livelihood approach for rural communities in South Africa. *African Journal of Development Studies* 11: 55. Available online: https://hdl.handle.net/10520/ejc-aa_affrika1_v11_n3_a3 (accessed on 1 March 2023).

Ministry of Agriculture, Livestock and Food Supply, Secretariat for Innovation, Rural Development and Irrigation Brasília. 2021. ABC + Plan for Adaptation and Low Carbon Emission in Agriculture Strategic Vision for a New Cycle. Available online: https://www.gov.br/agricultura/pt-br/assuntos/sustentabilidade/agricultura-de-baixa-emissao-de-carbono/publicacoes/abc-english.pdf (accessed on 1 March 2023).

Ministry of Food and Agriculture Republic of Ghana. 2018. Available online: https://mofa.gov.gh/site/images/pdf/AGRIC%2520IN%2520GHANA%2520F&F_2018.pdf (accessed on 15 April 2023).

Mishra, Aanchal. 2023. What to Know about India's Mid-Day Meal Scheme. The Borgen Project. Available online: https://borgenproject.org/indias-mid-day-meal-scheme/#:~:text=India%2520aimed%2520to%2520address%2520hunger,Poshan%2520Shakti%2520Nirman%2520Yojna%2520(PM (accessed on 16 April 2023).

Mueller, Bernd. 2019. Rural Youth Employment in Sub-Saharan Africa: Moving Away from Urban Myths and Towards Structural Policy Solutions. International Labor Organization. Available online: https://www.ilo.org/employment/Whatwedo/Publications/WCMS_790112/lang--en/index.htm (accessed on 15 April 2023).

Mukhopadhyay, Raj, Binoy Sarkar, Hanuman Sahay Jat, Parbodh Chander Sharma, and Nanthi S. Bolan. 2021. Soil salinity under climate change: Challenges for sustainable agriculture and food security. *Journal of Environmental Management* 280: 111736. [CrossRef] [PubMed]

National Nutrition Mission, India. n.d. Available online: https://www.india.gov.in/national-nutrition-mission?page=2 (accessed on 20 April 2023).

National Rural Livelihoods Mission—NRLM, India. n.d. Available online: https://aajeevika.gov.in/about/introduction (accessed on 25 March 2023).

Nilsson, Mans, Dave Griggs, and Martin Visbeck. 2016. Map the interactions between sustainable development goals. *Nature* 534: 320–23. [CrossRef] [PubMed]

Nilsson, Pia, Mikaela Backman, Lina Bjerke, and Aristide Maniriho. 2019. One cow per poor family: Effects on the growth of consumption and crop production. *World Development* 114: 1–12. [CrossRef]

Norway Government. 2021. The Government's Commitment to the Ocean and Ocean Industries. Available online: https://www.regjeringen.no/en/dokumenter/the-governments-commitment-to-the-ocean-and-ocean-industries/id2857445/?ch=3%20https://www.sciencedirect.com/science/article/pii/S0305750X18303565 (accessed on 13 February 2023).

Nutrition in Agriculture Program, Kenya—Nutrition Portal. n.d. Available online: https://www.nutritionhealth.or.ke/programmes/agri-food-and-nutrition-unit/ (accessed on 21 April 2023).

Organization of Women Honey Producers (WIHPA), India. n.d. Available online: https://www.wihpa.org (accessed on 11 March 2023).

Otto-Zimmermann, Konrad. 2012. From Rio to Rio+ 20: The changing role of local governments in the context of current global governance. *Local Environment* 17: 511–16. [CrossRef]

Pakkan, Sheeba, Christopher Sudhakar, Shubham Tripathi, and Mahabaleshwara Rao. 2023. A correlation study of sustainable development goal (SDG) interactions. *Quality & Quantity* 57: 1937–56. [CrossRef]

Pandiselvam, Ravi, Subhashini Sundaramoorthy, E. P. Banuu Priya, Anjineyulu Kothakota, S. V. Ramesh, and Shahir Sultan. 2019. Ozone based food preservation: A promising green technology for enhanced food safety. *Ozone: Science & Engineering* 41: 17–34. [CrossRef]

Parry, Bianca Rochelle, and Errolyn Gordon. 2021. The shadow pandemic: Inequitable gendered impacts of COVID-19 in South Africa. *Gender, Work & Organization* 28: 795–806. [CrossRef]

Malešević Perović, Lena, and Maja Mihaljević Kosor. 2020. The efficiency of universities in achieving sustainable development goals. *Amfiteatru Economic* 22: 516–32. [CrossRef]

Planting for Food & Jobs, Ghana. n.d. Available online: https://mofa.gov.gh/site/programmes/pfj#:~:text=Planting%2520for%2520Food%2520and%2520Jobs,market%2520and%2520also%2520provide%2520jobs (accessed on 1 April 2023).

Pradhan Mantri Garib Kalyan Yojana (PMGKY), India. n.d. Available online: https://www.ibef.org/government-schemes/pradhan-mantri-garib-kalyan-yojana (accessed on 1 April 2023).

Pradhan, Prajal, Luís Costa, Diego Rybski, Wolfgang Lucht, and Jürgen P. Kropp. 2017. A systematic study of sustainable development goal (SDG) interactions. *Earth's Future* 5: 1169–79. [CrossRef]

Productive Safety Net Program (PSNP), Ethiopia. n.d. Available online: https://essp.ifpri.info/productive-safety-net-program-psnp/ (accessed on 15 March 2023).

Reyers, Belinda, Mark Stafford-Smith, Karl-Heinz Erb, Robert J. Scholes, and Odirilwe Selomane. 2017. Essential variables help to focus sustainable development goals monitoring. *Current Opinion in Environmental Sustainability* 26: 97–105. [CrossRef]

Sachs, Jeffrey D., Guido Schmidt-Traub, Mariana Mazzucato, Dirk Messner, Nebojsa Nakicenovic, and Johan Rockström. 2019. Six transformations to achieve the sustainable development goals. *Nature Sustainability* 2: 805–14. [CrossRef]

Salasan, Cosmin, and Ioana Mihaela Balan. 2022. The environmentally acceptable damage and the future of the EU's rural development policy. In *Economics and Engineering of Unpredictable Events*. Edited by Taylor & Francis. London: CRC Grup Publisher Routledge, eBook ISBN 9781003123385. [CrossRef]

Schnetzer, Julian, and Lucie Pluschke. 2018. Solar-Powered Irrigation. Systems: A Clean-Energy, Low-Emission Option for Irrigation Development and Modernization. Available online: https://www.fao.org/3/bt437e/bt437e.pdf (accessed on 30 March 2023).

Sharma, Gagan Deep, Muhammad Ibrahim Shah, Umer Shahzad, Mansi Jain, and Ritika Chopra. 2021. Exploring the nexus between agriculture and greenhouse gas emissions in BIMSTEC region: The role of renewable energy and human capital as moderators. *Journal of Environmental Management* 297: 113316. [CrossRef]

Simonovic, Slobodan P. 2002. World water dynamics: Global modeling of water resources. *Journal of Environmental Management* 66: 249–67. [CrossRef]

Singh, Gerald G., Andrés M. Cisneros-Montemayor, Wilf Swartz, William Cheung, J. Adam Guy, Tiff-Annie Kenny, Chris J. McOwen, Rebecca Asch, Jan Laurens Geffert, Colette C. C. Wabnitz, and et al. 2018. A rapid assessment of co-benefits and trade-offs among Sustainable Development Goals. *Marine Policy* 93: 223–31. [CrossRef]

Smart Cities Mission India. n.d. Available online: https://smartcities.gov.in (accessed on 15 April 2023).

Spangenberg, Joachim H. 2017. Hot air or comprehensive progress? A critical assessment of the SDGs. *Sustainable Development* 25: 311–21. [CrossRef]

Stabile, Marcelo C. C., André L. Guimarães, Daniel S. Silva, Vivian Ribeiro, Marcia N. Macedo, Michael T. Coe, Erika Pinto, Paulo Moutinho, and Ane Alencar. 2020. Solving Brazil's land use puzzle: Increasing production and slowing Amazon deforestation. *Land Use Policy* 91: 104362. [CrossRef]

Stephens, Jennie C., Maria E. Hernandez, Mikael Román, Amanda C. Graham, and Roland W. Scholz. 2008. Higher education as a change agent for sustainability in different cultures and contexts. *International Journal of Sustainability in Higher Education* 9: 317–38. [CrossRef]

Sullivan, Kieran, Sebastian Thomas, and Michele Rosano. 2018. Using industrial ecology and strategic management concepts to pursue the sustainable development goals. *Journal of Cleaner Production* 174: 237–46. [CrossRef]

Supplemental Nutrition Assistance Program (SNAP) USA. n.d. Available online: https://www.fns.usda.gov/snap/supplemental-nutrition-assistance-program (accessed on 30 April 2023).

Swain, Ranjula Bali. 2018. A Critical Analysis of the Sustainable Development Goals. In *Handbook of Sustainability Science and Research*. Edited by W. Leal Filho. World Sustainability Series; Cham: Springer, pp. 341–55. [CrossRef]

The Lancet. 2019. The Global Syndemic of Obesity, Undernutrition, and Climate Change: The Lancet Commission Report. Available online: https://www.thelancet.com/journals/lancet/article/PIIS0140-6736(18)32822-8/fulltext (accessed on 13 April 2023).

The U.S. Government's Global Hunger & Food Security Initiative. Feed the Future. n.d. Available online: https://www.feedthefuture.gov (accessed on 23 April 2023).

Tremblay, David, François Fortier, Jean-François Boucher, Olivier Riffon, and Claude Villeneuve. 2020. Sustainable development goal interactions: An analysis based on the five pillars of the 2030 agenda. *Sustainable Development* 28: 1584–96. [CrossRef]

UN Climate Change. 2022. What Is the Triple Planetary Crisis? Available online: https://unfccc.int/blog/what-is-the-triple-planetary-crisis (accessed on 15 April 2023).

UN Global SDG Indicators Database. n.d. Available online: https://unstats.un.org/sdgs/indicators/database/ (accessed on 20 June 2023).

UN Goal 2: Zero Hunger. n.d. Available online: https://www.un.org/sustainabledevelopment/hunger/ (accessed on 17 April 2023).

UN Partnership for Action on Green Economy. 2022. Indonesia Launches Its Green Economy Index at G20. Available online: https://www.un-page.org/news/indonesia-launches-its-green-economy-index/ (accessed on 15 April 2023).

UN Progress towards the Sustainable Development Goals. n.d. Available online: https://unstats.un.org/sdgs/files/report/2017/secretary-general-sdg-report-2017--Statistical-Annex.pdf (accessed on 25 June 2023).

UN Resolution Adopted by the General Assembly on 6 July 2017 (A/RES/71/313). n.d. Available online: https://undocs.org/A/RES/71/313 (accessed on 1 June 2023).

UN The 17 Goals. n.d. Sustainable Development. Available online: https://sdgs.un.org/goals (accessed on 1 June 2023).

UN Transforming Our World: The 2030 Agenda for Sustainable Development. n.d. Available online: https://sdgs.un.org/2030agenda (accessed on 18 April 2023).

United States Department of State and the United States Executive Office of the President, Washington DC. 2021. The Long-Term Strategy of The United States. Pathways to Net-Zero Greenhouse Gas Emissions by 2050. Available online: https://www.whitehouse.gov/wp-content/uploads/2021/10/US-Long-Term-Strategy.pdf (accessed on 15 April 2023).

Vilalta, Josep M., Betts Alícia, and Gomez Victoria. 2018. Higher Education's role in the 2030 agenda: The why and how of GUNi's commitment to the SDGs. In *Sustainable Development Goals: Actors and Implementation. A Report from the International Conference (10–14)*. Barcelona: GUNi. Available online: https://www.acup.cat/sites/default/files/2018-06/Higher%2520Education%2527s%2520Role.pdf (accessed on 2 May 2023).

Vohra, K., and P. K. Saxena. 2022. Irrigation Growth in India—Prospects, Initiatives and Challenges. Available online: https://pmksy-mowr.nic.in/aibp-mis/Manual/Paper%2520on%2520IRRIGATION%2520GROWTH%2520IN%2520INDIA.pdf (accessed on 15 March 2023).

Ward, Hugh. 2006. International linkages and environmental sustainability: The effectiveness of the regime network. *Journal of Peace Research* 43: 149–66. [CrossRef]

WHO/UNICEF. 2008. *Progress on Drinking-Water and Sanitation: Special Focus on Sanitation*. Geneva: WHO/UNICEF. Available online: https://www.who.int/publications/i/item/9789280643138 (accessed on 1 April 2023).

Wong, Anson. 2021. The Interconnectedness of Sustainable Development Goals: Boom or Gloom. Earth Org. Available online: https://earth.org/the-interconnectedness-of-sustainable-development-goals/ (accessed on 2 April 2023).

World Bank. 2020. Promoting Sustainable Agriculture in Rio de Janeiro. Available online: https://www.worldbank.org/en/results/2020/10/23/promoting-sustainable-agriculture-in-rio-de-janeiro (accessed on 2 April 2023).

World Bank Open Data. n.d. Available online: https://data.worldbank.org/indicator/SL.AGR.EMPL.ZS?locations=ZG (accessed on 15 March 2023).

World Food Programme. 2020. 5 Facts about Food Waste and Hunger. Available online: https://www.wfp.org/stories/5-facts-about-food-waste-and-hunger (accessed on 15 March 2023).

World Food Programme. Hunger Map. 2022. Available online: https://www.fao.org/interactive/state-of-food-security-nutrition/2-1-1/en/ (accessed on 15 March 2023).

World Health Organization. 2016. *World Health Statistics 2016 [OP]: Monitoring Health for the Sustainable Development Goals (SDGs)*; Geneva: World Health Organization. Available online: https://books.google.ro/books?hl=en&lr=&id=-A4LDgAAQBAJ&oi=fnd&pg=PP1&ots=dcqfZPdlwy&sig=wkq_oFjW7fVG9txp6DQGTDKAXys&redir_esc=y#v=onepage&q&f=false (accessed on 10 March 2023).

The Power of Social Capital to Address Structural Factors of Hunger

Gian L. Nicolay

1. Introduction to the Problem of Hunger as a Problem of Lack of Theory

Hunger has not yet been defeated, despite the considerable progress made in science, technology, and business development in the last 70 years (IAASTD 2009). In this essay, I argue that the local social capital, particularly in the Global South, is a neglected factor by science, economy, and politics, explaining the slow progress made by humanity towards achieving food security, resilient food systems, and the transition towards "Zero Hunger". Hunger creates social unrest and revolution. From a human rights perspective, it remains a scandal in this world of plenty and food production exceeding the needs of the world population, at least when caused by structural factors that political measures could correct it (Orr and Lubbock 1953; de Castro 1952; Ziegler 2010). Looking from the risk perspective (Centeno et al. 2015), the globalized food system will create, in the future, new threats of massive hunger due to uninspected breakdowns of trade, transport, social order, and high-impact plant or animal diseases. The distance between vulnerability and catastrophe is decreasing with the growth of interdependencies and the ongoing reduction in family farms worldwide (Erenstein et al. 2021), or at least the reduction in farm size in low-income countries (Lowder et al. 2016).

The debates related to hunger and food system transformation reflect the complexities of the task more and more. Science is not only seen as an objective observer, but is called upon to participate in public debates (Caron et al. 2021). However, science, apart from being considered embedded in society as a functioning system (Luhmann 1995), is not simply providing voices and forces or speaking with one voice. Politicians can pick out the research data and messages which are the most convenient to them. Nevertheless, the political arena remains the most relevant one in addressing the hunger problem, as issues of inequality of rights and income, poverty, conflicts and wars, climate change, and the often-inefficient food supply chains remain the critical drivers of hunger and are under the responsibility of politics dealing with public affairs and freedom. The latest Food Security and Nutrition report by the FAO et al. (2021) states, amongst other things, that:

- New projections confirm that hunger will not be eradicated by 2030 unless bold actions are taken to accelerate progress, especially actions to address inequality in access to food.
- Close to 12 percent of the global population was severely food insecure in 2020, representing 928 million people—148 million more than in 2019.
- The high cost of healthy diets coupled with persistent high levels of income inequality puts healthy diets out of reach for around 3 billion people, especially the poor, in every region of the world in 2019. This number is slightly less than in 2017 and will likely increase in most regions in 2020 due to the COVID-19 pandemic.
- Most children with malnutrition live in Africa and Asia. These regions account for more than nine out of ten of all children with stunting, more than nine out of ten children with wasting and more than seven out of ten children who are affected by overweight worldwide.
- Conflict, climate variability and extremes, and economic slowdowns and downturns (now exacerbated by the COVID-19 pandemic) are major drivers of food insecurity and malnutrition that continue to increase in both frequency and intensity, and are occurring more frequently in combination.
- Drivers that are external (e.g., conflicts or climate shocks) and internal (e.g., low productivity and inefficient food supply chains) to food systems are pushing up the cost of nutritious foods which, combined with low incomes, are increasing the unaffordability of healthy diets. (FAO et al. 2021).

Let us read these statements by the FAO through the lens focused on local social capital. We have a more concrete view of why more weight should be put on the local level and to better understand why humanity is not progressing on this question of ending hunger, but rather has been turning in circles for over 40 years at least. We will take a closer look at the critical keywords "bold actions", "inequality in access to food", "the poor", "economic slowdowns and downturns", "low productivity", and finally, "inefficient food supplies". In the following chapters, we will try to focus not only on relating these factors in a systemic way to social capital, but also to substantiate them with experiences made in concrete territories, communities, and local food systems in various regions of Africa over the last 15 years since the 2007/08 food crises (Sommerville et al. 2014).

My 40 years of experience in the field of food and agriculture and rural development worldwide taught me that the hunger issue in contemporary society is too complex to be dealt with meaningfully by science and as a knowledge issue alone, as it is a foremost practical and political reality, with its weakest links being the local level of social capital in poor countries. Solutions always require simultaneous action and agency on all four levels, i.e., local, national, regional and global, in order to address the hunger problem in a systematic, inclusive and effective way. Knowledge has to be complemented by appropriate action and agency (as the capacity to act). To underestimate the complexity of the task by leaving out a factor such as social capital would lead to failure. Additionally, over-rating the complexity would be fatal. We must manage this paradox, that the hunger question is complex and simple. It is simple in the sense that we find the ways out when becoming concrete, knowing the local conditions, accepting power (or control) relations, and progressing step-by-step using our pragmatic and ethic-based attitudes in given social networks and through dialogue and openness, adding to the required social capital of the given local or national food system. Human thinking is a proven way to manage paradoxes, difficulties, and crises. From Aristoteles to Ernst Bloch (Zimmermann 2016), we have plenty of wisdom, methods, and principles at our disposal. Ernst Blochs' famous Principle of Hope books should always remind scientists and practitioners to make the best use of thinking, arguing, debating, and acting as a continuous flow. This, briefly, means that we have to better understand the concept and reality behind "social capital". However, first, let us look at hunger's generally agreed upon or at least mentioned structural drivers.

2. Generally Mentioned Structural Drivers of Hunger

Hunger is, fortunately, the exception to the rule in contemporary times, thanks to outstanding achievements in agriculture, agronomy, technology, and in building fine-tuned institutional systems over the "civilized" planet, linking all or most countries and regions and providing food assistance in case of large emergencies. However, the exception has still touched, over the last 20 to 50 years, about 10% of humans. A scientific consensus is that meeting the food demand by 2050 by applying sustainable food production would be one of humanity's most significant challenges (Cassman and Grassini 2020). It is therefore vital to understand the commonly agreed upon critical drivers of hunger, capable of interrupting the target of solving the hunger problem in the coming 30 years, assuming that no major technological breakthrough nor change on the demand side will happen during this time. We will briefly present poverty, wars and pandemics, climate extremes, gender, age and race, societal divisions, and capitalist economies as commonly mentioned drivers. This list is of course not complete. However, I intend to open the view of the large specter of hunger, a phenomenon that involves many aspects of society, history, and natural phenomena of our planet.

2.1. Poverty

Poverty is the most significant risk factor for hunger in all cases, whether the affected people are producers or consumers (Cooper et al. 2021). Poverty often means not having a voice and not always being represented in important events. The former FAO DG stated that very clearly:

> What makes hunger a very complex political problem is that the hungry are not represented. I never saw a union association that represents the malnourished . . . Most people who face hunger nowadays are not in this situation due to a lack of food produced but because they don't have money to buy it. So, give them money or the resources to gain access to food. It is a simple formula. The best would be to increase employment and the minimum wage paid to a level that could allow workers access to a healthy diet. And for those who can't be employed for different reasons, provide them a minimum subsidy through cash transfer programs, as we did in Brazil's Zero Hunger program. It is that simple: there is no miracle! (Jose Graziano da Silva, IPS Interview 23 September 2021 (Wise 2021))

The poor consumers and the working class in rich countries are in a different position compared to food-insecure people and households in poor or low-income countries; they can, in most cases, count on food stamps, programs, and various institutions amid plenty.

2.2. Wars and Pandemics

Wars, including civil wars, are ideal for interrupting food production and increasing the probability of hunger. In 2020, almost all low- and middle-income countries were affected by pandemic-induced economic downturns. The number of undernourished people was more than five times greater than the highest increase in undernourishment in the last two decades. When other drivers also affected those countries, particularly climate-related disasters, conflict, or a combination, the most significant increase in undernourishment was seen in Africa, followed by Asia (FAO et al. 2021).

2.3. Climate Variability and Extremes

Climate variability and extremes affect land degradation, yields, animal health, and food security (IPCC 2020). However, they simultaneously affect communities and institutions and hence the food systems. Areas receiving less rain, less appropriate rain patterns, and higher temperatures may reduce their cropping and herding areas and will be affected with less production and decreased food security. Dependency on food aid thus increases. The WFP (2021) states:

> Even before the COVID-19 pandemic, the world was not on track to end hunger and all forms of malnutrition by 2030. In 2020, hunger and malnutrition shot up in absolute and proportional terms, largely perpetuated by the socio-economic effects of COVID-19. However, unlike COVID-19, there is no vaccine to protect vulnerable communities worldwide from the worsening climate crisis. By 2050, the risk of hunger and malnutrition could rise by 20 percent if the global community fails to act now to mitigate and prevent the adverse effects of climate change. (WFP 2021, p. 1)

We will not go into further detail here, assuming that this issue is well-known or broadly accepted by most readers.

2.4. Gender, Age, and Race

A recent report summarizes the situation for women involved in food systems as follows:

> After 12 years, global food security governance is highly fragmented, with the power of a small number of actors increasing dramatically. Those actors include major multinational corporations, the World Bank and the IMF and the G7 governments. The voices of the people who have been left food-insecure are seldom heard in policy discussions. Funding targeted at women in agriculture is insignificant compared with other official funding, and this public disinvestment opens the door to other actors, such as multinational companies, which have taken a "business as usual" approach and make gender equality in agriculture a low priority at best. Especially in light of climate change and increased conflicts, failing to address the structural causes of the food price crisis has put women even more at risk on all dimensions of food security. (Botreau and Cohen 2020, p. 104)

It is globally recognized that achieving gender equality and empowering women is an absolute precondition to breaking the cycle of poverty and hunger and achieving the 17 Sustainable Development Goals (FAO 2021a). The FAO advises ensuring that gender equality issues are included in all spheres and that there is monitoring and reporting progress at the country level.

Does race matter? In 1964, Boyd Orr[1] and Lubbock wrote in "The White man's Dilemma" in codified style about the prevailing importance of racist prejudices:

> However, if all the people in the world had had environmental conditions which would enable them to attain their full inherited capacity for physical and mental ability, there would be little, if any, the difference between the ability of men of different races. (Op. cit, p. 73)

2.5. Societal Contradictions and Divisions and Poor Education

The Agenda 21 organized by UNCED in 1992 already tried to effectively address poverty and hunger with a global and non-binding international initiative. This ambitious program was soon attacked by an anti-Agenda 21 movement and was replaced 24 years later by the more ambitious SDGs. The governments failed in most cases to implement their commitments (DESA 2012). Bureaucracies and loss of national sovereignty were common reasons for collectively attacking and boycotting international plans to address poverty and hunger. The contradictions of the capitalist development divide the most relevant actors of society, from the poor farmers and laborers to the shareholders of multinational companies. A lack of debates and poor education contribute to undermining institutions and rules such as ignoring international biodiversity and climate change protocols, particularly those on food (Fakhri 2021). Other consequences of such societal divides and poor education are fast-growing meat consumption and cereals produced for agro-fuel, both increasing hunger conditions for the poor.

2.6. Capitalist Economies Including Land Acquisitions by Non-Farmers

The literature on the risks related to globalized food systems is still growing, but already reveals important general features (Holt-Giménez and Altieri 2012). More specific criticism is related to local biophysical and social sub-systems (de Raymond et al. 2021) and their complexities (Liu et al. 2007). Financial markets, operating in an unregulated space with futures and other instruments, and the often-missing relation between local and subnational markets (e.g., grain in the Sahel) from international prices are topics emerging in the research of risks related to food systems. Clashes between global investors and local communities are happening and programmed, when divergent interests (shareholder profit/private investments replacing sound

---

[1] He was the first DG of the FAO (1945–1948) and winner of the Nobel Peace prize for his research on nutrition in 1949.

national policies of rural development vs. community-driven and food sovereignty) collide (McMichael 2012), fail to communicate (Luhmann 1995) and increase the vulnerability of food-insecure countries (de Raymond et al. 2021). Smallholders, the laborers, rural communities in low- and mid-income countries (LMIC) on one side, and the poor consumers depending on cheap (but unhealthy) food leading to obesity and health problems on the other side, pay the price. Risks related to system breakdowns are increasing, as previously resilient and rather autonomous food networks at the local level have rapidly disappeared worldwide. Often discussed under the names of "rural exodus" or "migration", but well-explained with the concepts of exploitation and suppression, this phenomenon is highly related with the concept of local social capital—or alternatively related to the more general concept of a social sub-system. For Luhmann (1997), exploitation, suppression and any other moral reasons do not explain the destiny of the poor and the hungry today, as they are "just outdated mythologies". The question is rather one of inclusion or exclusion into the world's society and its highly differentiated and complex functional systems (such as economy, law, media, politics).

3. Social Capital as an Underestimated Parameter for Reducing Hunger

We may consider social capital featuring social organization such as networks, norms, and social trust, that facilitate coordination and cooperation for mutual benefit (Putnam 1995, p. 67). Social capital is rarely used as a factor or parameter in explaining food insecurity and hunger.

The notion of capital emerged in economics with Adam Smith and evolved over time by differentiating into natural, financial, human, social and institutional capital (Piazza-Georgi 2002). In the 1960sm it was also commonly agreed within agricultural economics, that human capital—however, not yet the related social capital—is a prerequisite for economic growth, particularly in low-income countries (Melton 1965). Since then, independently from economics, sociology took up the concepts of human and social capital (Portes 1998), but corrected the focus from the mainly individual human capital concept towards social structures (such as networks) and inter-personal agency, emphasizing the social dimension. What is social capital from the sociology perspective? The answer depends on the applied concepts, theories and narratives. I use the tradition and knowledge emanating mainly from contemporary sociology, particularly based on Pierce, Weber, Elias, Luhmann, Bourdieu, and notably, Harrison White (2008). They all try to understand issues of social organization and chaos, identities and actors, power and control, institutions and styles[2], mores and ideas/ideologies, culture and history, from the local or micro level up to a global society level. Most of the sociologists mentioned are adhering somehow to a relational perspective, which emphasizes networks with processes such as interactions, social ties, cooperation and conversations as the central material of the social. In the last 60 years, a huge collection of literature in both economics and sociology has emerged, providing evidence on the critical impact of social capital—or the quality of the population as it is also called by Schultz (1981)—on poverty alleviation and addressing hunger in LMIC.

If we refer to the famous definition of social capital provided by Bourdieu (1980, 1986), which according to some scholars (Julien 2014; Portes 1998) is still one of the most appropriate and precise descriptions of this concept emanating from sociology, then we can define it, as follows, as a working concept:

> Social capital is the aggregate of the actual or potential resources which are linked to possession of a durable network of more or less institutionalized relationships of mutual acquaintance and recognition—or in other words, to membership in a group. Each group or community has a certain amount of social wealth in various forms- participation and bonds to larger public goods through social relations in networks; access to information and knowledge; trust amongst its members; inclusiveness-, which can be used to different degrees by its members and so further accumulate or consume its resources. (slightly modified by the author from Bourdieu 1986)

This concept has therefore a double importance in the hunger debate. For the individual actor, this "aggregate of resources" or networks (of social relationships) constitute a determining point of the departure for agency, which yields reproductive benefits such as access to new connections and networks, markets, information, and credit through gained confidence, influence, status and trust (Nicolay 2017). The individual farmer receives agency in order to better organize in collective action, take up opportunities and invest (Schultz 1981) and build social structures or institutions in solving his or her problems (Nicolay 2016). From the collective (or social) side, and here comes the second importance of social capital, it provides a pool of relations which helps in maintaining and reproducing its agency as a collective or corporate agent or identity. Social capital understood as a complex of networks intermingled with self-reproducing context and providing nested structures (such as

[2] In the sense of White (2008), i.e., the rhythm of life, whereas identities are like melodies.

communities) within a social system reduces its vulnerability against "attacks" of control by competing collective or individual agents (White 2008). Examples of social capital phenomena are: (a) A local group of smallholders that increases its chances to fight against intentions of foreign corporate organizations to take control of their land use by new formats of internal organization, or; (b) Organic cotton unions who defend their price interests and selling conditions within the national cotton network often dominated by traditional and corporate interests (Nicolay 2019). Social capital is therefore always embedded and intermingled with social networks (or social systems). It enhances the probability of creating effective forms of organizations, such as committees or innovation platforms, which are instrumental in both innovation adaptation and collective local problem solving (Nicolay 2016).

The Negligence of (Local) Social Capital as a Parameter of Hunger

The specification "local" shall mean that I focus the discussion of the social capital on territorial or geographic levels, where food and fiber are produced through labor. Neglecting local social capital can have fatal consequences for society (Helbing 2013), including its food systems. We have seen that social components matter in the food and hunger nexus, but that both science and practice struggle to make them visible and understand the reality of social relations. Social order/chaos are undervalued, and the importance of atomistic/individual actors overstated. This may be primarily explained by the lack of philosophical and sociological thinking[3] and the dominance of "common sense" or non-reflective language. Examples of social capital having a key role in succeeding in sustainable intensification (as an important element of food system in LMIC) and improved economic performance are extension systems, agencies linking farmers to markets and external agencies, innovation platforms (Adekunle and Fatunbi 2012), farmer field schools, cooperatives (Biresaw 2019; Muriqi et al. 2021) and business groups (Pretty et al. 2011). However, what if such projects building on enhancing social capital (or social infrastructure according to Pretty) remain exceptions and are not taken up by policy measures at national level?

---

3 Luhmann remarked: People see humans, science (sociology) sees systems".

4. A Critique of Agroecology and Its Ideological Fight against Agribusiness

The concept and approach of agroecology has gained a strong momentum and even support from the FAO since the last food crises in 2009 as a way to contribute to food security. The main features according to the FAO (2021b) are:

- Agroecology is a holistic and integrated approach that simultaneously applies ecological and social concepts and principles to the design and management of sustainable agriculture and food systems. It seeks to optimize the interactions between plants, animals, humans and the environment while also addressing the need for socially equitable food systems within which people can exercise choice over what they eat and how and where it is produced.
- It now represents a transdisciplinary field that includes food systems' ecological, socio-cultural, technological, economic and political dimensions, from production to consumption.
- It is no longer possible to look at food, livelihoods, health and the management of natural resources separately. Embracing systems thinking through holistic approaches is needed to address these complex and interdependent challenges. The fundamental connection between people and the planet, with sustainable agriculture and food systems, is at the heart of the 2030 Agenda for Sustainable Development, which stresses the urgent need to take concerted action and pursue policies directed at transformational change.
- Ending poverty, achieving zero hunger while ensuring inclusive growth, and sustainably managing the planet's natural resources, all in the context of climate change and biodiversity loss, will only be possible through holistic and integrated approaches that respect human rights.
- Agroecology is based on bottom-up and territorial processes, helping to deliver contextualized solutions to local problems with people at the center. There is no single way to apply agroecological approaches—it depends on local contexts, constraints and opportunities but there are common principles that have been articulated in the framework of the 10 Elements[4] of Agroecology.

Agroecology is therefore mainly determined by territorial, local and human-nature-related networks and links, including the concept of social capital at various levels such as the household, village/neighborhood and local markets):

---

4 The ten elements of agroecology are: diversity, co-creation and sharing of knowledge, synergies, efficiency, recycling, resilience, human and social values, culture and food tradition, responsible governance and circular and solidarity economy (see Figure 1)

> Agroecology places a strong emphasis on human and social values, such as dignity, equity, inclusion and justice all contributing to the improved livelihoods dimension of the SDGs. It puts the aspirations and needs of those who produce, distribute and consume food at the heart of food systems. By building autonomy and adaptive capacities to manage their agro-ecosystems, agroecological approaches empower people and communities to overcome poverty, hunger, and malnutrition, while promoting human rights, such as the right to food, and stewardship of the environment so that future generations can also live in prosperity (Human and social values). (FAO 2021b; Knowledge hub, p. 1)

Interestingly, the economic dimension is not emphasized, nor are ecology or the social dimension. Should large-scale farms and agribusiness not consider and apply agroecology principles as well? However, more severe is the rather moral claim (aspirations, needs) of the here abstract workers and farmers (emphasis on human and social values) instead of the real and concrete requirements in the form of people's needs and social capital.

The local networks at the farm and landscape levels are often poorly linked to the higher-level institutions dealing with governance and control, and in most cases, are even conflictual. The intended or proclaimed support for the transition to sustainable food and agricultural systems towards food security following agroecological concepts and practices is not happening (IPES-Food 2020). This structural feature—the critical relation between the two levels—often fails exactly due to precisely the social capital at the local level. By visualizing the position of the ten agroecology elements (see below) and differentiating on the axes of natural/social and local/high-level, the neglect of "social capital" (a part of the negligence of the economic dimension) in the current design of agroecology becomes apparent (see Figure 1).

Even if agroecology, compared with its counter concept of industrial agriculture, is relatively strongly emphasizing the social aspects within agriculture and food systems, the conceptualization of social capital and the real needs of the farmers and laborers remains weak.

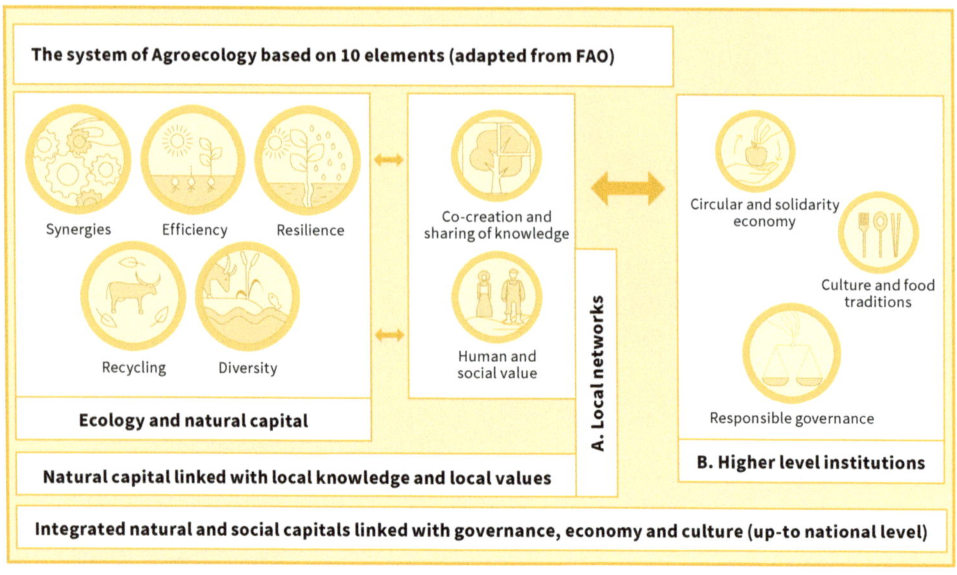

Figure 1. The positioning of social capital in the agroecology system as promoted by the FAO. Source: Figure adapted by the author.

Ideology Disputes between Agroecology Movements and Agribusiness

Power relations between producers (labor) and buyers (financial capital) at local to global levels require more attention again. This is about controlling the direction, benefits cashed in by individuals, groups and social classes, as well about the outcomes and impacts of food systems on nature, society and economics. The complex of intermingling networks of buyers–producers—networks constituting markets (White 2008)—forge the multiple food systems worldwide. Gliessman and Montenegro (2021) complain, for example, that the organizers of the UNFSS[5] relegate agroecology to further promoting competing technologies such as genetic engineering, digital frameworks, and big data, undermining the interests of (family) farms in favor of corporate interests. However, researchers have tried to mediate here by stating, that "socio-economic researchers need to suggest inclusive ways to transform the more than 400 million smallholder farms worldwide. They must identify pathways out of inequitable and unfair arrangements over land, credit and labor, and empower the rights of women and youth" (von Braun et al. 2021).

5 UN Food Security Summit organized by UN in September 2021.

A concerned voice from research regarding the current role of the private sector and investors in the global food summit governance comes from Covic et al. (2021):

> Further, the agenda and discussions at the UNFSS (UN Food Security Summit) need to be driven by LMICs and other voices that are often given a back seat, including citizens, consumers, and other civil society actors. While many experts have been called upon or have simply stepped up to organize and be present at the UNFSS, others may purposely stay away due to legitimacy concerns. For some constituencies, the involvement of trans-and multinational food businesses in the UNFSS is problematic, given their dominance over global and national food systems and a history of multiple instances of bad practice. On the other hand, driven by consumer demands and more sustainability-focused investors, there are significant changes in some parts of the private sector towards environmental sustainability. (p. 3)

From the institutional side, various UN rapporteurs have been struggling for over 15 years to have the "Rights to food" respected and to convince public and private actors to better respect this right in the plans and policies at national level. They lament that issues of power, participation, and basic rights to food remain unresolved. Interestingly, the rapporteurs clearly take a position for agroecology[6] as a way to address the problems of food insecurity and hunger (Fakhri 2021) and consider it as an alternative to industrial agriculture. This is interesting, as I see no reason why industrial farming practices should not be forced to apply the principles, assuring a path towards ecology and social fairness for all operators. Another strong voice from the producers' side is Via Campesina[7]. They have advocated for over 20 years for the concept of food sovereignty, implicitly emphasizing social capital and agency at national and local levels as critical requirements for abolishing hunger.

5. The Morphological Look to Overcome Ideological Barriers

Ideological disruptions and fights at the global to local level have of course important implications on networks in food systems, but they are inevitable. Can we mediate to a certain degree such ideological impasses? I propose here a way to apply the morphological analyses. Ritchey (1998) argues that the morphological analysis,

---

6 It is "often more productive than intensive industrial techniques" (Fakhri 2021, p. 7)

7 See its Manifesto for the future of the planet on https://viacampesina.org/en/food-sovereignty-a-manifesto-for-the-future-of-our-planet-la-via-campesina/ (accessed on 16 October 2021).

as invented by Zwicky[8] in around 1950, is an appropriate and ingenious method to deal with complex policy issues. I would add that it could also address blocked dialogues due to strong ideological differences, as we observe between two forms of commonly perceived food regimes: (i) agribusiness (or "industrial agriculture"), which overemphasizes the economic dimension, and (ii) the agroecology movement[9], which considers itself a reaction to agribusiness. The method is based on building all possible combinations or figurations within a fictitious (but plausible) solution space:

> An alternative to formal (mathematical) methods and causal modelling is a form of non- quantified modelling relying on judgmental processes and internal consistency, rather than causality. Causal modelling, when applicable, can- and should- be used as an aid to judgement. However, at a certain level of complexity (e.g., at the social, political and cognitive level), judgement must often be used, and worked with, more or less directly. The question is: How can judgmental processes be put on a sound scientific basis? Sets of non-quantified conditions can be synthesized into well-defined relationships or configurations representing solution spaces. In this context, there is no fundamental difference between quantified and non-quantified modelling. (Ritchey 1998, p. 2)

In Table 1, I try such a modelling with defined relationships of configurations representing a solution space. I propose ten parameters and allow three options for each (A, B, C), also representing "ideotypical" food regimes. By allowing three options, a new space appears against the dual perspective. The column "B" represents the new option. Agroecology cannot be the new option, as we see no solution outside the restricted (only smallholders, only local circles, only movements) agroecological production regime (see Figure 1) in order to be sustainable. Agroecology positions are mostly captured with option A (and partially B), whereby agribusiness (or industrial agriculture) is mainly covered under option C. Under B fall the middle options, such as middle-sized family-based agribusiness or well-off modern farms and households with sufficient labor and land capital. All main forms of food systems worldwide can be captured in this way and can subsequently be discussed. Critical solutions are marked in Table 1 with the dark background in grey.

---

8 Swiss astrophysicist and engineer.

9 However, international organisations such as the FAO and EU are also supporting agroecology, but not in the same way as various social movements.

Table 1. Morphological box for food and agriculture systems, based on 10 parameters providing 30 combinations (shaded combinations considered critical).

	Parameter	Regime A	Regime B	Regime C
1	Production unit	Family farm (poor)	Family and/or cooperative farm (supported)	Corporate or state farm
2	Production method (bottom-line for A and B: agroecology*)	Local inputs dominating	Mixed	Monocultures (high-tech, capital-intensive)
3	Type of markets	Nested/local**	Mixed	Global
4	Social distance between producers and end-consumers	Small	Medium	Big
5	Price fixation	Producer–consumer agreements	Mixed	Downstream VC actors
6	Policy space	National	all levels	International
7	Justice/human rights- and community-based (of food system)	Yes	Partial	No
8	Role of lead scientific disciplines	Natural and economics	Natural, socio-economics, ethics	Private sector research (mostly natural and economics)
9	Value-addition location (mainly processing of food)	Local	Subnational and larger areas	Global (rich countries)
10	Multifunctionality of agriculture systems	Yes (>4 functions)	Partial (2–3 functions)	No (only economic functions)
	Critical options (in grey fields)	*2*	*0*	*5*

Source: Table by author. *See definition by the FAO. **See a definition in Fakhri 2021 on "territorial markets" (p. 20).

The production unit is a key and first parameter to distinguish "agroecology as a movement" from "industrial agriculture". In Table 1, "Agroecology as a movement" is in the solution space 1AB and "Industrial agriculture" in 1BC, as industrial farming is also possible on non-capitalist-owned farms of middle size (1B). The solution space 1B (supported family farms) can be both agroecological and industrial. These two polarizing concepts are revealed here as concepts rather than realities[10]. In reality, it has to be seen case-by-case.

The morphological view further provides 7 critical constellations out of 30:

- The poor family farms (1A) have economic and often environmental deficiencies and are hardly socio-economically viable (often not attractive enough to young farmers to take over from their parents).
- The capital and high-tech production method (2C) fostering monoculture (on tropical soils) has either social or environmental deficiencies (or both).
- The dominance of price fixation by downstream value chain actors (5C) leads to the unjust distribution of created wealth and socio-economic disruptions.
- The lack of respect and institutionalization of human rights (7C) impedes the emergence of autopoietic local structures able to cope with context and required investments.
- Social and societal dimensions including social capital and politics are not sufficiently considered in option A, where the link between local food production and world consumption needs are neglected (8A).
- The lack of local and subnational value addition in favor of urban elites at the national level in the Global South and the rich countries (9C) prevents the emergence of required wealth accumulation in rural LMIC and ultimately perpetuates the conditions for "structural" hunger.
- Agriculture systems only designed and used as an economic function (10C) will run into conflicts with society and nature (see FAO et al. 2021).

We can maintain the two dominant regimes of food systems in the Global South (observation of literature and of public debates; see also the cited Gliessman, Covic and Fakhri): (i) the (small-scale) farmer-based and (ii) agribusiness-based systems dominated by the corporate and capitalist-motivated[11] networks. Agroecology is often attributed to the small-scale-based regime (see Figure 1 with the element of circular and solidary economy). However, from the agronomic side, there is no reason that large-scale corporate and state farms could not fulfil most of the agroecology

---

[10] This is not surprising for practitioners and researchers working on organic agriculture, which can work on all three types of production units (A, B, C).

[11] Increasing numbers are urban-based families and groups involved in agribusiness, operating through employees and tenants. In Africa since the 1990s.

criteria. The morphological box (Table 1) informs us about five critical features emanating from agribusiness regimes and only two from the (poor, small-scale) farmer regime. Policies should therefore mainly concentrate on adjusting the five critical features of agribusiness and ensure that the agroecological principles are applied here. Secondly, we believe that policies should better address the destiny of poor farming households—or better prevent their farms from going bankrupt or just being given away. The policy should ensure their inclusion and further existence and role in the food system or in any other economic sector.

The bottom line for assessing food regimes is sustainability. We can now describe the solution space in tabular form as self-perceived by the two food regimes AEM (agroecology movement) or Option A, BUZ (agribusiness) or Option C, or finally of SDG or option B (Table 2). Both mainly discussed food regimes (A, B) show weak elements in the chain as compared to the ideal of the SDG (Sustainable Development Goals) and the Option B regime. The following pattern of farmer-based or mixed forms are considered sustainable options and compatible with SDG 1 (poverty) and 2 (hunger), thus providing a solution space.

Table 2. Solution space for three food regimes.

Regime Type	Regime Type or Case Description	Solution Pattern Based on the Ten Parameters	N $_{(critical)}$
AEM/regime A (agroecology movement)	Farmer, ideal from within	1AB-2A-3AB-4AB-5AB-6ABC-7A-8B-9A-10A	2
BUZ/regime C (agribusiness)	Agribusiness, ideal from within (corporate food regime)	1C-2C-3C-4BC-5C-6A-7C-8C-9C-10C	5
SDG/regime B	Sustainable according to SDG	1BC-2B-3BC-4AB-5B-6ABC-7AB-8B-9ABC-10AB	0

Shaded/yellow combinations are considered critical from the outside/author observer's perspective (Perceptions depend on interests and background. Therefore, there is a need for debates to find agreements). Source: Table by author.

The conclusion from the morphological analyses and the solution space is simple:

- Agro-industry (or agro-business) and agroecology (or family farming), seen as either a movement or network, can be clearly differentiated. However, the respective concepts or ideal figurations behind them are not clearly separable and should be seen as the two poles between a continuum

- A more significant part of the concrete agribusiness-driven food systems is not (yet) based on agroecological principles, is not inclusive, reduces the chances for local social capital formation and does not contribute to poverty alleviation (no multi-functionality)
- The farmer-based regime has its weakness in its large fraction of very resource-poor farms (or households) and efficiency problems (constraining investment options)
- Price fixation agency and local value addition are two key features in which the two regimes A and C differentiate in the political–economic dimension. As agribusiness is in most cases also present downstream the value chain, they can fetch better profit than farms that rarely have negotiation power, i.e., only in specific markets and if well-organized (e.g., in federations of cooperatives) at the national level. These two features require social capital from the involved primary producers.

6. Ways and Tactics to Strengthen Local Social Networks in Food Systems

Innovation platforms, going beyond the scope of a single value chain, are institutions with the potential to increase local value addition and contribute to creating wealth and increasing food security. They provide ideal starting points to strengthen local capacities and networks and thus provide concrete and practical ways to create social capital. A practical application requires theoretical and practical skills, leadership, and proactive engagement of various stakeholders going beyond the involved value chains. In addition, it requires financial partners. Fostering agency, empowerment, and the growth of social capital as an impact target would increase the chances of strengthening the agency of farmer figurations. Innovation platforms can be organized by combining 1–3 value chains plus additional core themes such as soil fertility improvement and maintaining the social coherence within a given area (commune, district, etc.).

6.1. A Better Link between Local and National Actors and Programs

One example to show the need for a better connection between local and national scales is the very dynamic land market in SSA. Land prices have skyrocketed recently (Jayne et al. 2021). What will be the implications on food security, the accessibility of land for young farmers, and social equity and economic growth?

Another example is the need to better harness the current and coming opportunities to use weather forecast data to inform farmers and herders on time (Nakalembe et al. 2021). This is only possible if the links between national and local actors are assured.

A third example is a need for a balanced and efficient food supply-chain growth at the local level (Reardon et al. 2021), assuring a healthy supply of processed food ideally produced locally and in the country. For such systems to be built, a fine-tuning of activities and interactions between cooperatives, policymakers, local entrepreneurs, and food suppliers and processors is called for.

Farmer organizations must strengthen alliances with strategic partners and emancipate themselves from the narrow national corset.

I consider the following actor groups as crucial allies to boost social capital within food systems: technology providers and financial partners, the judiciary, social scientists, consumers, and the international community.

Technology providers and entrepreneurs are prime allies and are often in short supply, as the markets are not attractive enough and rarely promoted during the critical stages by the public sector (Mazzucato 2021). Financial partners and investors are part of the required ecology of partners and networks to promote social capital-based networks with viable but fair business plans.

The example I would like to present for the importance of financial partners and investors relates again to West Africa, particularly the organic cotton sector. I have been personally involved in its promotion since 1998 through the work of the Swiss NGO Helvetas, where we, as a group of practitioners, organic farming promoters, entrepreneurs, and scientists, created the West Africa Organic Cotton Coalition (CCBE) in 2017. The CCBE aims to address eight critical challenges for the 2.5 million cotton producers of West Africa with an innovative approach based on the concept of social capital: (1) low productivity of cotton production; (2) poverty and food insecurity of producers; (3) soil degradation and declining yields; (4) uncertainty about how programs are coping with climate change; (5) unemployment and low labor wages; (6) health problems due to pesticide application in the cotton fields; (7) global textile companies having difficulties in finding organic cotton lint in the world market, and; (8) social unrest (Nicolay 2019). It took us over three years to win a financial partner (GIZ) to set the coalition in motion and start reaching a tangible impact, covering four producing countries (Burkina Faso, Benin, Mali, and Senegal). This example shows that in complex systems such as those of cotton and textile, which are highly differentiated and globalized, social capital and networks require assistance from the financial side if they cannot wait to generate their financial capital.

6.2. Politics, the Legal Framework, and the Judiciary

Sustainable intensification often depends on public support to address market limitations such as low farm prices and to adhere to environmentally sound practices (Silva et al. 2021). One example from Mali exemplifies the importance of politics and the legal and judiciary framework when aiming to upscale successful local initiatives related to food security and climate change adaptation (Bautze et al. 2018). Upscaling often requires substantial financial and human resources and a proper institutional framework. In the case of Mali, evolving between 2012 and 2018, the more the difficulties in implementing projects grew, the more hierarchies of the state from regions to national or central levels were involved. One explanation for this is that the drivers' social capital was insufficient and unable to mobilize new influential members from the top level, based on the national figuration and networks controlling more considerable funds, influences, and decision-makers. The collaborative efforts stopped abruptly between the regional (Sikasso) and national levels. Nevertheless, the step from grassroots/commune to the regional level succeed within a few months, involving local administrative, legal, and judiciary actors.

6.3. Social Scientists Applying Holistic and Inclusive Concepts

We can define a holistic concept if it addresses a perceived societal problem such as poverty and hunger or soil fertility degradation and biodiversity loss in a given area in all three dimensions of the SDG. Social scientists are skilled in selecting and using the required tools and approaches and mobilizing and uniting the various specialists and stakeholders to design solutions and viable pathways. By including representatives from farmer communities and groups with generally weak (unheard) voices and low social capital, new constellations can be created between researchers, farmers, and local and national decision-makers. Such innovative processes can lead to further research findings, encouraged farmer networks, and stimulated local economies and communities. The state can bring in its agency by mobilizing resources and facilitating processes enhancing food security. Research findings such as a better knowledge of the dynamics of farm class behavior—moving towards professionalizing or resilience or instead towards decapitalizing and withdrawing from the sector—allow the local stakeholders to better target their advice and investments. Thanks to the recognition and better understanding of markets, social networks and social dynamics on the ground can emerge (Frossard et al. 2017; Nicolay et al. 2020). Hence, the critical social capital is built and can trigger the envisaged socio-economic and ecologic development.

6.4. Consumers, Citizens and the International Community

Consumers and citizens, mediated by honest media and political activists, are more relevant, as large amounts of food are distributed, processed and traded in globalized networks. Consumer choice can make huge differences in which value chains to consider and what to buy. Such options finally decide which producers or production units will benefit. Consumers' choices often depend on how they are informed (and educated) by the media. More solidarity is required between the working class (hoping for relatively low food prices) and the farmers (aiming for higher food prices). This alliance is highly complex and challenging, as it must go beyond national borders.

The International community constituted by the UN system, state interventions, civil society, and fairness-based responsible businesses, does not directly impact the local level. Nevertheless, its policies, measures, actions, and investments play a critical role in shaping the socio-ecological landscape at the local level. Depending on the context, there are six pathways to follow toward food systems' transformation: (1) integrating humanitarian, development, and peacebuilding policies in conflict-affected areas; (2) scaling up climate resilience across food systems; (3) strengthening the resilience of the most vulnerable to economic adversity; (4) intervening along the food supply chains to lower the cost of nutritious foods; (5) tackling poverty and structural inequalities, ensuring interventions are pro-poor and inclusive, and; (6) strengthening food environments and changing consumer behavior to promote dietary patterns with positive impacts on human health and the environment (FAO et al. 2021). These pathways must be dealt with through coherent progressive politics; if not, they fall apart and remain ineffective.

7. Conclusions

Structural hunger can be defeated. Taking note of all the factors, using our current knowledge, and fixing the aim with agency, hunger can be beaten within 20 to 30 years. However, systems approaches based on comprehensive philosophies and narratives are needed to build coherent portfolios of policies, investments, and legislation and enable win–win solutions while managing trade-offs. These include various approaches: territorial, ecosystems, indigenous peoples' food systems approaches, and interventions that systemically address protracted crisis conditions (FAO et al. 2021). Additionally, the conscious and professional use of social capital theories and practices are required to assure that the policies, investments, legislation, and approaches reach the people where the critical actions happen—at the local level (village, neighborhood, commune, district). Innovation platforms have been presented as a possible starting point for creating social capital.

More policy support (Cooper et al. 2021), better links with health and nutrition (Welch and Graham 2000), and prolonged standing advice are also required. More diverse and bottom-up-based food security and governance approaches (Sommerville et al. 2014), and at the same time a reduction in the fragility of the trade-based global food system (Puma et al. 2015), should be discussed more intensely after the experiences with COVID-19. When improving social capital, the critical items mentioned by the FAO et al. (2021) can be addressed: bold actions, ecological crises, inequality in access to food, the needs of the poor, low productivity, inadequate food supplies, and finally, hunger.

The good news is that the pathway toward ending hunger is not different from the pathway towards achieving SDGs, including the creation of more inclusive societies (or, with Luhmann, to create an inclusive world society) or the pathway to mitigating the disastrous impacts of the climate chaos. These three enormous challenges—the fight against hunger and for inclusiveness, sustainability, and climate change—will probably occupy the public discourse in the next 10 to 20 years. Understanding that they have similar root causes enhances the chances of finding collective solutions at the world level.

This essay is the result of over 40 years of thought and work with rural communities, the UN system (FAO), INGO work in Africa, the private sector, the working class, and a research institution. Its messages and reflections may be heard or not. It is based on the tradition of philosophical thought aiming to make this world a better place for all humans by reducing hunger. It is all in our hands, hearts, and brains. Finally, as Ernst Bloch reminds us, in learning again together and in tolerance but determination, hope.

Funding: This research is supported by the Swiss National Science Foundation (SNSF) as a follow-up of the ORM4Soil and Yamsys projects under the r4d programme (Swiss Programme for Research on Global Issues for Development).

Conflicts of Interest: The author declares no conflict of interest.

References

Adekunle, A. A., and Abiodun Oluwole Fatunbi. 2012. Approaches for Setting-up Multi-Stakeholder Platforms for Agricultural Research and Development. *World Applied Sciences Journal* 16: 981–88.

Bautze, Lin, Gian L. Nicolay, Matthias Meier, Andreas Gattinger, and Adrian Muller. 2018. Climate Change Adaptation Through Science-Farmer-Policy Dialogue in Mali. In *Handbook of Climate Change Resilience*. Edited by W. Leal Filho. Cham: Springer International Publishing, pp. 1–15.

Botreau, Hélène, and Marc J. Cohen. 2020. Chapter Two—Gender inequality and food insecurity: A dozen years after the food price crisis, rural women still bear the brunt of poverty and hunger. In *Advances in Food Security and Sustainability*. Edited by Marc J. Cohen. Amsterdam: Elsevier Science, pp. 53–117.

Biresaw, Amelework. 2019. Role of Social Capital in Agricultural Producers' Cooperatives for Commercialization: In the Case of Ethiopia. Available online: https://ssrn.com/abstract=3393202 (accessed on 17 August 2021).

Bourdieu, Pierre. 1980. Le capital social. *Actes de la Recherche en Sciences Sociales* 2–3. [CrossRef]

Bourdieu, Pierre. 1986. The forms of capital. In *Handbook of Theory and Research for the Sociology of Education*. Edited by J. Richardson. New York: Greenwood, pp. 241–58.

Caron, Patrick, Martin Van Ittersum, Tessa Avermaete, Gianluca Brunori, Jessica Fanzo, Ken Giller, Etienne Hainzelin, John Ingramg, Lise Korstenh, Yves Martin-Préveli, and et al. 2021. Statement based on the 4 TH international conference on global food security—December 2020: Challenges for a disruptive research Agenda. *Global Food Security* 30: 100554. [CrossRef]

Cassman, Kenneth G., and Patricio Grassini. 2020. A global perspective on sustainable intensification research. *Nature Sustainability* 3: 262–68. [CrossRef]

Centeno, Miguel A., Manish Nag, Thayer S. Patterson, Andrew Shaver, and A. Jason Windawi. 2015. The Emergence of Global Systemic Risk. *Annual Review of Sociology* 41: 65–85. [CrossRef]

Cooper, Matthew, Benjamin Müller, Carlo Cafiero, Juan Carlos Laso Bayas, Jesús Crespo Cuaresma, and Homi Kharas. 2021. Monitoring and projecting global hunger: Are we on track? *Global Food Security* 30: 100568. [CrossRef]

Covic, Namukolo, Achim Dobermann, Jessica Fanzo, Spencer Henson, Mario Herrero, Prabhu Pingali, and Steve Staal. 2021. All hat and no cattle: Accountability following the UN food systems summit. *Global Food Security* 30: 100569. [CrossRef]

de Castro, Josué. 1952. *The Geography of Hunger*. Little: Brown.

de Raymond, Antoine Bernard, Arlène Alpha, Tamara Ben-Ari, Benoît Daviron, Thomas Nesme, and Gilles Tétart. 2021. Systemic risk and food security. Emerging trends and future avenues for research. *Global Food Security* 29: 100547. [CrossRef]

DESA. 2012. Review of Implementation of Agenda 21 and the Rio Principles. Study prepared by the Stakeholder Forum for a Sustainable Future. Synthesis. Available online: https://sustainabledevelopment.un.org/content/documents/641Synthesis_report_Web.pdf (accessed on 28 September 2021).

Erenstein, Olaf, Jordan Chamberlin, and Kai Sonder. 2021. Farms worldwide: 2020 and 2030 outlook. *Outlook on Agriculture* 50: 003072702110255. [CrossRef]

Fakhri, Michael. 2021. Report on the Right to Food. Transmitted to the General Assembly in Accordance with Assembly Resoulution 75/179. Available online: https://digitallibrary.un.org/record/3936776?ln=en (accessed on 23 September 2021).

FAO, IFAD, UNICEF, WFP, and WHO. 2021. The State of Food Security and Nutrition in the World. Transforming Food Systems for Food Security, Improved Nutrition and Affordable Healthy Diets for All. Available online: http://www.fao.org/documents/card/en/c/cb4474en (accessed on 29 June 2021).

FAO. 2021a. *Achieving Gender Equality and Women's Empowerment in Agriculture and Food Systems—A Handbook for Gender Focal Points*. Rome: FAO.

FAO. 2021b. *Agroecology Knowledge Hub*. Rome: FAO. Available online: http://www.fao.org/agroecology/overview/en/ (accessed on 17 August 2021).

Frossard, Emmanuel, Beatrice A. Aighewi, Sévérin Aké, Dominique Barjolle, Philipp Baumann, Thomas Bernet, Daouda Dao, Lucien N. Diby, Anne Floquet, Valerie K. Hgaza, and et al. 2017. The Challenge of Improving Soil Fertility in Yam Cropping Systems of West Africa. *Frontiers in Plant Science* 8: 1953. [CrossRef]

Gliessman, Steve, and Maywa de Wit Montenegro. 2021. Agroecology at the UN food systems summit. *Agroecology and Sustainable Food Systems* 45: 1417–21. [CrossRef]

Helbing, Dirk. 2013. Globally networked risks and how to respond. *Nature* 497: 51–59. [CrossRef] [PubMed]

Holt-Giménez, Eric, and Miguel A. Altieri. 2012. Agroecology, Food Sovereignty, and the New Green Revolution. *Agroecology and Sustainable Food Systems* 37: 90–102. [CrossRef]

IAASTD. 2009. *Agriculture at a Crossroads. International Assessment of Agricultural Knowledge, Science and Technology for Development (IAASTD): Global Report*. Colombo: International Water Management Institute.

IPCC. 2020. *Climate Change and Land. An IPCC Special Report on Climate Change, Desertification, Land Degradation, Sustainable Land Management, Food Security, and Greenhouse Gas Fluxes in Terrestrial Ecosystems. Summary for Policymakers*. Geneva: IPCC.

IPES-Food. 2020. Money Flows: What Is Holding Back Investment in Agroecological Research for Africa? Biovision Foundation for Ecological Development & International Panel of Experts on Sustainable Food Systems. Available online: www.agroecology-pool.org/MoneyFlowsReport (accessed on 27 September 2021).

Jayne, T. S., Jordan Chamberlin, Stein Holden, Hosaena Ghebru, Jacob Ricker-Gilbert, and Frank Place. 2021. Rising land commodification in sub-Saharan Africa: Reconciling the diverse narratives. *Global Food Security* 30: 100565. [CrossRef]

Julien, Chris. 2014. Bourdieu, Social Capital and Online Interaction. *Sociology* 49: 356–73. [CrossRef]

Liu, Jianguo, Thomas Dietz, Stephen R. Carpenter, Marina Alberti, Carl Folke, Emilio Moran, Alice N. Pell, Peter Deadman, Timothy Kratz, Jane Lubchenco, and et al. 2007. Complexity of Coupled Human and Natural Systems. *Science* 317: 1513–16. [CrossRef]

Lowder, Sarah K., Jakob Skoet, and Terri Raney. 2016. The Number, Size, and Distribution of Farms, Smallholder Farms, and Family Farms Worldwide. *World Development* 87: 16–29. [CrossRef]

Luhmann, Niklas. 1995. *Social Systems*. Stanford: Stanford University Press.

Luhmann, Niklas. 1997. Globalization or World society: How to conceive of modern society? *International Review of Sociology* 7: 67–79. [CrossRef]

Mazzucato, Mariana. 2021. *Mission Economy: A Moonshot Guide to Changing Capitalism*. London: Penguin Books Limited.

McMichael, Philip. 2012. The Land Grab and Corporate Food Regime Restructuring. *Journal of Peasant Studies* 39: 681–701. [CrossRef]

Melton, R. B. 1965. Schultz's Theory of "Human Capital". *The Southwestern Social Science Quarterly* 46: 264–72.

Muriqi, Shyhrete, Zsolt Baranyai, and Maria Fekete-Farkas. 2021. Comparative analysis of cooperative & non-cooperative farmers in Kosovo. *Economics and Sociology* 14: 242–63. [CrossRef]

Nakalembe, Catherine, Inbal Becker-Reshef, Rogerio Bonifacio, Guangxiao Hu, Michael Laurence Humber, Christina Jade Justice, John Keniston, Kenneth Mwagi, Felix Rembold, Shraddhanand Shukla, and et al. 2021. A review of satellite-based global agricultural monitoring systems available for Africa. *Global Food Security* 29: 100543. [CrossRef]

Nicolay, Gian L. 2016. Theory-Based Innovation Platform Management. A Contribution of Sociology to Agriculture Research and Development. Paper presented at the 12th European IFSA Symposium, Newport, UK, July 12–15; Available online: http://www.harper-adams.ac.uk/events/ifsa-conference/papers/2/2.5%20nicolay.pdf (accessed on 27 June 2021).

Nicolay, Gian Linard, Louis Chikopela, Boubacar Diarra, and Andreas Fliessbach. 2020. Modelling farm behaviour on soil fertility with the policy variable: A case from Africa. *Organic Agriculture* 10: 43–56. [CrossRef]

Nicolay, Gian Linard. 2017. Why is Africa struggling with organic farming? A methodological contribution from sociology. Paper presented at the Organic World Congress 2017, Delhi, India, November 9–11.

Nicolay, Gian Linard. 2019. Understanding and Changing Farming, Food & Fiber Systems. The Organic Cotton Case in Mali and West Africa. *Open Agriculture* 4: 86–97.

Orr, Lord Boyd, and David Lubbock. 1953. *The White Man's Dilemma*. London: George Allen and Unwin, Ltd.

Piazza-Georgi, Barbara. 2002. The role of human and social capital in growth: Extending our understanding. *Cambridge Journal of Economics* 26: 461–79. [CrossRef]

Portes, Alejandro. 1998. Social Capital: Its Origins and Applications in Modern Sociology. *Annual Review of Sociology* 24: 43–67. [CrossRef]

Pretty, Jules, Camilla Toulmin, and Stella Williams. 2011. Sustainable intensification in African agriculture. *International Journal of Agricultural Sustainability* 9: 5–24. [CrossRef]

Puma, Michael J., Satyajit Bose, So Young Chon, and Benjamin I. Cook. 2015. Assessing the evolving fragility of the global food system. *Environmental Research Letters* 10: 024007. [CrossRef]

Putnam, Robert D. 1995. Bowling Alone: America's Declining Social Capital. *Journal of Democracy* 6: 65–78. [CrossRef]

Reardon, Thomas, David Tschirley, Lenis Saweda O. Liverpool-Tasie, Titus Awokuse, Jessica Fanzo, Bart Minten, Rob Vos, Michael Dolislager, Christine Sauer, Rahul Dhar, and et al. 2021. The processed food revolution in African food systems and the double burden of malnutrition. *Global Food Security* 28: 100466. [CrossRef]

Ritchey, Tom. 1998. Fritz Zwicky, Morphologie and Policy Analysis. Available online: https://citeseerx.ist.psu.edu/viewdoc/download?doi=10.1.1.457.9454&rep=rep1&type=pdf (accessed on 24 September 2021).

Schultz, Theodore W., ed. 1981. *Investing in People. Economics of Population Quality*. Berkeley: University of California Press.

Silva, Joao Vasco, Pytrik Reidsma, Frédéric Baudron, Alice G. Laborte, Ken E. Giller, and Martin K. van Ittersum. 2021. How sustainable is sustainable intensification? Assessing yield gaps at field and farm level across the globe. *Global Food Security* 30: 100552. [CrossRef]

Sommerville, Melanie, Jamey Essex, and Philippe Le Billon. 2014. The 'Global Food Crisis' and the Geopolitics of Food Security. *Geopolitics* 19: 239–65. [CrossRef]

von Braun, Joachim, Kaosar Afsana, Louise O. Fresco, and Mohamed Hassan. 2021. Food systems: Seven priorities to end hunger and protect the planet. *Nature* 597: 28–30. [CrossRef] [PubMed]

Welch, Ross M., and Robin D. Graham. 2000. A new Paradigm for World Agriculture: Productive, Sustainable, Nutritious, Healthful Food Systems. *Food and Nutrition Bulletin* 21: 361–66. [CrossRef]

WFP. 2021. *Climate Crisis and Malnutrition A Case for Acting Now*. Rome: World Food Programme.

Wise, Timothy A. 2021. Transforming Global Food Systems. Jose Graziano da Silva on the Path to Zero Hunger. *IPS News*, April 6. Available online: https://www.ipsnews.net/2021/04/transforming-global-food-systemsjose-graziano-da-silva-path-zero-hunger/ (accessed on 27 September 2022).

White, Harrison C. 2008. *Identity and Control: How Social Formations Emerge*. Princeton: Princeton University Press.

Ziegler, Jean. 2010. *Betting on Famine: Why the World Still Goes Hungry*. New York: New Press.

Zimmermann, Rainer E. 2016. *Ernst Bloch: Das Prinzip Hoffnung*. Berlin: De Gruyter.

The Implications of Agroecology and Conventional Agriculture for Food Security and the Environment in Africa

Terence Epule Epule and Abdelghani Chehbouni

1. Introduction

Agroecology is a form of agriculture that is dependent on natural systems of cultivation and minimum external inputs. Some of its properties include aspects of agriculture that are reliant on natural systems and very minimal external inputs, such as green manure, compost, crop rotation, and biological pest control. Agroecology also includes the use of pesticides and fertilizers generated from natural avenues, such as pyrethrin from flowers and bone meal from animals (Badgley et al. 2007). On the other hand, conventional agriculture is any form of agriculture that is based essentially on external inputs, such as inorganic fertilizers, pesticides, and herbicides, inter alia (Badgley et al. 2007; Seufert et al. 2012). Therefore, the key differences between agroecology and conventional agriculture are anchored on the fact that conventional farming is based on inorganic or synthetic fertilizers, pesticides, and plant growth catalysts, such as antibiotics and hormones, while agroecology is based on natural inputs (Badgley et al. 2007; Lindell et al. 2010a). In a nutshell, the major differences between agroecology and conventional agriculture are based on decentralization, independence, community, harmony with nature, diversity, and restraint (Lindell et al. 2010a, 2010b) (Figure 1).

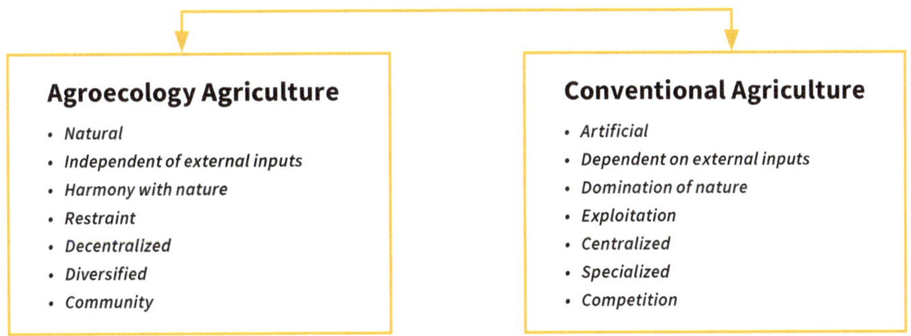

Figure 1. The differences between agroecology and conventional agriculture. Source: Figure by authors.

The global carrying capacity of the Earth is being affected by rapid population growth throughout the world. The ability of the Earth to feed a projected population of about 9 to 10 billion people by 2050 is being compromised by the galloping world population (Badgley et al. 2007). Thus, there is a pressing need to enhance global production and at the same time ensure that the environmental footprints of current and emerging production systems do not perturbate the promises of intergenerational sustainability (Seufert et al. 2012). Agroecology responds to the twin challenges aimed at producing food while at the same time reducing the environmental footprints of production activities. Despite the benefits of agroecology, the system has been criticized for being unbale to feed the global population when compared to conventional agriculture (Seufert et al. 2012). To ensure better yields and for farming systems under agroecology to level up and be parallel with current global conventional systems of production, it is necessary to bring more land into cultivation. The idea of expanding agricultural land through extensive farming heralds evidence-based repercussions on the environment through effects such as the loss of forests, biodiversity degradation, and inadequate organically acceptable farming procedures that can produce enough food without compromising the environmental strengths of agroecology (Trewavas 2001; Seufert et al. 2012; Epule 2019).

The prospects of brining more land into production in Africa is diminishing, as land expansion is becoming a limitation (Morris et al. 2007). A study by Badgley et al. (2007, p. 86) observes that agroecology has the potential of contributing quite substantially to the global food supply, while at the same time protecting the environment. The latter observation has been criticized by Seufert et al. (2012) on the grounds that 1. they included agroecology yields from farms with inputs of large amounts of nitrogen from manure; 2. they included less representative low conventional yields in their comparison; 3. they failed to consider yield reductions over time due to rotations with non-food crops; 4. the double counting of high agroecology yields; and 5. the extensive use of unverifiable data from the grey literature.

With all these criticisms in mind, it is a complex process to predict if agroecology will maintain its promise of producing more food and protecting the environment. Unfortunately, the Seufert et al. (2012) paper cannot be used to conclude this regarding Africa, as it fails to use data from the latter. Its conclusions cannot be generalized because it is more germane to say there are insufficient studies dwelling on this topic on an African scale except for (Epule 2019).

2. Methodology

The data used in this chapter are based on a review of the peer-reviewed and grey literature. A search for suitable publications was performed in the following search engines: Google Scholar, Scopus, Institute of Scientific Information (ISI), and Scientific Citation Index (SCI) Web of Science. It is believed that the data culled from these search engines are representative, as ground truthing did not result in any new studies. This work considers suitable literature in both French and English, and a total of 49 peer-reviewed and grey literature studies were included in this review. Of these 49 publications, only 6 were published by authors with affiliations in African universities or organizations. From these publications, 30 focused on agroecology while 19 focused on conventional agriculture. The time span of the papers selected was open, but all the studies included in the review had a focus on Africa or some African country. The search was conducted using keywords such as agroecology and conventional agriculture in Africa, benefits and challenges of agroecology and conventional agriculture in Africa, and food security implications of agroecology and conventional agriculture in Africa. From the resulting database, key themes were identified based on the objectives of this chapter.

3. Theoretical Foundations of Agroecology and Conventional Agriculture

Agricultural systems can be balkanized into two broad categories, which include agroecology and conventional agriculture. These farming systems represent two broad agricultural systems. For a long time, unless an agricultural system is being productive in terms of yields, using more inorganic fertilizers and pesticides in addition to resulting in cutting up more forest in favor of farmland expansion, it is said to be facing a decline. This emphasis on yields is often at the detriment of the environment, and a society is said to face a decline if it cannot keep up with yield demands at all costs and at the expense of environmental degradation, deforestation, and greenhouse gas emissions, inter alia. For a very long time, agricultural systems around the world have been based on conventional agriculture. This dominant or traditional system of farming places emphasis on the following aspects: 1. Inputs of inorganic fertilizers and hybrid sowing materials or seeds (Morris et al. 2007; Matson et al. 1997; Epule et al. 2012). 2. Mechanization or dependence on agricultural machinery. 3. Monoculture or the cultivation of single crops. 4. The commercialization of production through a huge market orientation as well as development and integration into a global economy (Figure 2a). Conventional agriculture has been associated with environmental degradation, with the Global South being more vulnerable due to poverty, accessibility, and general low adaptive capacities (Lindell et al. 2010a, 2010b; Matson et al. 1997; Rosegrant and Cline 2003; Snapp et al. 2010; Hossain and Singh 2000; Reid et al. 2003; Valenzuela 2016).

On the other hand, agroecology has been proposed as an alternative to conventional agriculture for the following benefits: 1. It enhances crop production. 2. It protects the environment by reducing environmental footprints due to its minimal dependence on external inputs. 3. It is easily accessible by poor farmers in the Global South since it is less dependent on synthetic external inputs. The key components of agroecology include the following: 1. It is based on the use of natural nutrient cycling with little or no synthetic substances (Badgley et al. 2007; Snapp et al. 2010; Kerr et al. 2007). 2. It ensures improvements in yield through the following options: 1. Nutrient cycling through natural processes and the accumulation of organic matter. 2. Natural control of pests and diseases using relationships such as predator–prey strategies rather than the use of pesticides, as well as the use of animal wastes as pesticides. 3. The conservation of environmental resources such as energy, soil, biodiversity, and water. 4. The enhancement of biodiversity, synergies, and interactions (Figure 2b). The fundamental principle behind this paradigm is anchored on resolving the debates around the dilemma related to feeding mankind while at the same time ensuring that the environment is protected.

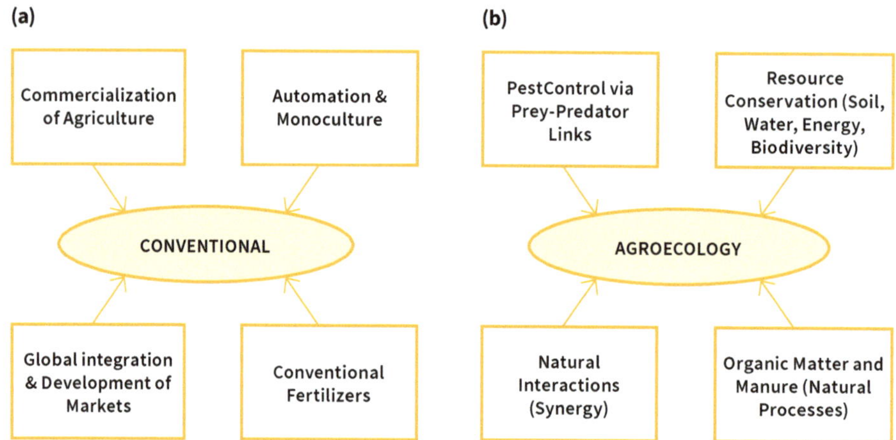

Figure 2. The paradigms of (**a**) the conventional paradigm and (**b**) agroecology. Source: Authors' compilation based on data from Epule et al. (2015).

4. Benefits and Challenges of Agroecology and Conventional Agriculture in Africa

4.1. *Agroecology*

One of the key benefits of agroecology is that most of its options are freely available. This aspect is of great importance in the African context because most agricultural production is in the hands of smallholder peasant farmers who often do not have enough money to invest in costly conventional agricultural inputs (Pichot et al. 1981; Bationo and Mokwunye 1991; Rosegrant and Svendsen 1993; Bado et al. 1997; Altieri 2009). With agroecology, manure and compost can easily be obtained from household waste, while animal droppings and urine can be used as fertilizers and pesticides (Snapp et al. 2010; Rosegrant and Svendsen 1993). Additionally, predator–prey relationships can be used to combat pests through the entomology technique of breeding insects that prey on other harmful pests on farms. All these options are freely available at little or no cost (Epule et al. 2012, 2015).

Furthermore, agroecology options are always environmentally friendly or compatible with natural environmental systems (Snapp et al. 2010; Rosegrant and Svendsen 1993). This is evident as composts, manure, and other natural inputs that make up agroecology do not pollute the soil, nearby streams, and other water resources (Pichot et al. 1981; Bationo and Mokwunye 1991; Bado et al. 1997, 2007). Instead, they help to enhance soil aeration and social organic carbon (Epule et al. 2012, 2015).

Again, as seen in the literature review, when agroecology is properly valorized it leads to improved yields. In fact, the low-level capabilities of agroecology in Africa are currently tied to the lack of valorization (Pichot et al. 1981; Bationo and Mokwunye 1991; Bado et al. 1997, 2007). This can be achieved through sensitization and the creation of pilot demonstration farms (Epule et al. 2012, 2015).

Finally, agroecology is open to the diversification of agriculture through either mixed cropping or mixed farming (Pichot et al. 1981; Bationo and Mokwunye 1991; Bado et al. 1997, 2007). In addition to the fact that this diversifies farmers' income when some crops are not successful, it also provides an opportunity for the introduction of livestock (Snapp et al. 2010; Rosegrant and Cline 2003). In fact, with livestock, agroecology can be enhanced because the waste from the animals can be used as manure and to control pests at the farm level, while some of the thrash from the crops can be decomposed to provide manure as well as be used as food for the animals (Epule et al. 2012, 2015).

One of the main disadvantages of agroecology is that it is often dependent on the incorporation of more land into agricultural systems (Epule et al. 2012, 2015). This is disadvantageous, because when more land is cultivated, it is likely that more trees will be cut down (Pichot et al. 1981; Bationo and Mokwunye 1991; Bado et al. 1997, 2007). In fact, the prospects of agroecology are highly limited in the context of farmland expansion, as there is currently evidence that there is less land for the continuous expansion of farms. Continuous deforestation exposes the land to various types of erosion and long-term soil degradation (Epule et al. 2012, 2015). This is the case in West, Central, and East Africa, as farmers in these regions depend mostly on farmland expansion to increase yields.

In addition, agroecology is currently not well-developed in Africa due to problems with the valorization of the systems (Pichot et al. 1981; Bationo and Mokwunye 1991; Bado et al. 1997, 2007). Most farmers cannot afford inorganics and end up farming under peasant conditions without exploring the elements of agroecology. This low level of valorization accounts for the often-low commercialization orientation of this type of agriculture (Snapp et al. 2010; Rosegrant and Cline 2003). If most countries in Africa valorize this system, as is the case in Malawi, it is likely that it might be able to provide yields that are parallel to those of conventional farms (Epule et al. 2012, 2015).

4.2. Conventional Agriculture

One of the benefits of conventional agriculture in Africa is that it is less dependent on land expansion (Snapp et al. 2010; Rosegrant and Cline 2003). In contrast to agroecology, which is often extensive and land-dependent, conventional agriculture is usually intensive, focusing on intensive inputs of fertilizers, machines, and pesticides, inter alia (Pichot et al. 1981; Bationo and Mokwunye 1991; Bado et al. 1997, 2007). The advantage of this is that there is less deforestation and soil degradation caused by different types of soil erosion (Epule et al. 2012, 2015).

Secondly, another advantage of conventional agriculture is that it is highly market-dependent and pro-commercialization (Snapp et al. 2010; Rosegrant and Cline 2003). In other words, most production in highly intensive farms and its associated investments are often aimed at a huge market orientation (Pichot et al. 1981; Bationo and Mokwunye 1991; Bado et al. 1997, 2007). The benefits of commercial agriculture are that it provides an income for famers and makes agriculture a primary industry that operates beyond subsistence (Epule et al. 2012).

In addition, conventional agriculture is often intensive and mechanized. This means that farmers under this system more often use machines, which go a long way to enhance production (Snapp et al. 2010; Rosegrant and Cline 2003). This is in contrast to most agroecology-related approaches, which often depend on natural systems and human labor (Pichot et al. 1981; Bationo and Mokwunye 1991; Bado et al. 1997, 2007). The incorporation of machines into conventional agriculture is an illustration of the valorization of conventional agriculture and, evidently, the occurrence of higher crop yields under this system of cultivation (Epule et al. 2012).

In terms of the challenges, it is important to note that the inputs into conventional farming systems are often not compatible with natural environmental systems (Snapp et al. 2010; Rosegrant and Cline 2003). This is seen in the case of inorganic fertilizers, which may enhance yields but, on the other hand, pollute streams and other underground water resources if they are not properly managed (Pichot et al. 1981; Bationo and Mokwunye 1991; Bado et al. 1997, 2007). Recent research has even shown that inorganic fertilizers and pesticides, in addition to other hormones found in the latter, have been described as partly responsible for several illnesses, such as cancer (Epule et al. 2012).

Furthermore, conventional agriculture is often monocultural, meaning that often only a single crop is cultivated (Snapp et al. 2010; Rosegrant and Cline 2003). The limitation of this is that, when there are crop failures, farmers might have many difficulties in recovering from the loss as they might not have suitable alternatives (Pichot et al. 1981; Bationo and Mokwunye 1991; Bado et al. 1997, 2007). Additionally, monocultural farms are often more susceptible to changes in climate and are therefore more easily affected when bad climatic conditions occur (Epule et al. 2012). This is evident as the farmers involved might not have suitable alternatives, as they depend on a single crop.

Finally, conventional agriculture is often characterized by inorganic fertilizers, pesticides, and fungicides, and is hugely machine-intensive (Pichot et al. 1981; Bationo and Mokwunye 1991; Bado et al. 1997, 2007). These key components of agroecology are expensive and not readily accessible in most African countries, as most farmers are poor and unable to afford them (Snapp et al. 2010; Rosegrant and Cline 2003). In cases where governments have invested to enhance access, rife rates of corruption have resulted in having these resources end up in the wrong hands (Epule et al. 2012).

5. Agroecology and Conventional Agriculture: Food Security Implications in Africa

According to the Malthusian population perspective, global food production was going to be unable to sustain the galloping world population. As a result of this, the debate is now around the potential of agroecology- and conventional-farming-related approaches to meet global food needs in general and especially food needs in the Global South. The initial hypothesis is that conventional agriculture can produce higher yields, in contrast to agroecology (Badgley et al. 2007; Epule et al. 2015). Indeed, conventional agriculture has witnessed major technological advancements in the last six decades. These changes were necessitated by the pressure imposed by the doubling of the human population in the past 40 years and have resulted in the production of more food. However, the problem now is one associated with the distribution of the food gains as well as the toll that these practices exert on the environment (Badgley et al. 2007; Abedi et al. 2010). Within this distribution chain, it can be observed that the Global South is still faced with problems of rife food insecurity and associated increased vulnerability to climate stressors. With an increase in the global consumption of meat and a decline in grain production, it is more important than ever to tilt production towards more sustainable methods. This is because conventional systems do not only degrade the environment through a strong dependence on external inputs but have also not been able to solve the problems of food insecurity in many parts of the world, including Africa (Altieri 1995, 2002; Uphoff 2003).

However, it has been argued that the alternative of conventional agriculture, which is agroecology, is still not capable of ensuring production to attain levels that equate to those of conventional production. In addition, agroecology requires more land and renders this approach unsustainable, as it often results in deforestation and consequent land degradation through different forms of erosion (Stoop et al. 2002; Tilman et al. 2002; Bumb and Groot 2004). Badgley et al. (2007) verified the criticisms against agroecology through two models of food production under agroecology conditions. The first model used agroecology: non-agroecology yield ratios based on developed world studies. It is argued that when converted into agroecology, low-intensity agriculture, which is a phenomenon of most of the developing world, will produce similar or slightly lower yields as obtained in the developed world. The subsequent model used other yield ratios derived from the developed world as well as yield ratios culled from studies on the developing world and applied them to food production in the developing world. The latter study found that agroecology can indeed contribute to feed the current and future world populations with little or no environmental concerns. These results are said to under-represent the real yields in agroecology, as several farms were reported for individual crops because many agroecology systems use multiple cropping approaches, within which the total output is often higher when compared to single crops (Badgley et al. 2007).

Since more research has focused on conventional agriculture, there are more opportunities to increase agroecology, as much remains unknown. Therefore, if the same emphasis that has been placed on conventional agriculture is placed on agroecology, agroecology is likely to result in additional yields. Furthermore, the yield per unit seems to be higher for agroecology-based systems of production for smaller than bigger farms in both the developed and developing world. This means that an increase in the number of small farms will further enhance global food production (Seufert et al. 2012; Bado et al. 2007). These results do not imply that yields under agroecology exceed yields under conventional yields but suggest rather that under certain conditions agroecology has the potential to enhance crop yields (Badgley et al. 2007).

The results that show that agroecology can feed the world's population have been debunked and criticized by Seufert et al. (2012). These criticisms are anchored on the justification that the authors of those findings used data from crops that were not purely under agroecology, and therefore were erroneous in their comparison of yields. In their study, Seufert et al. (2012) performed a comprehensive meta-analysis of the literature on agroecology and conventional agriculture. The principal findings showed that conventional agriculture is more capable of feeding the world's population when compared to agroecology (Seufert et al. 2012).

There is a variation in agroecology yields according to different crop types. For example, yields under agroecology for oil seeds and fruits show a very small but significant difference when compared to conventional yields. Additionally, vegetables and cereals witnessed significantly lower yields than conventional crops (Seufert et al. 2012; Bationo and Mokwunye 1991; Bado et al. 2007; Tilman et al. 2002; Bumb and Groot 2004). From a time, perspective, agroecology-related yields are usually low during the first year and gradually increase with time due to enhancements in soil fertility. Under rain-fed conditions, agroecology yields are higher than those under irrigation. Considering a global perspective, agroecology yields are said to be higher in the developed world than in the developing world. However, the general conclusion of the Seufert et al. (2012) paper is that agroecology-based systems of agriculture generally have lower yields when compared to conventional yields.

In Africa, there are several studies that have assessed the performance of agroecology and conventional agriculture in the context of their impacts on crop yields (Altieri 1995, 2002; Pretty 1995; Gliessman et al. 1998; Kerr et al. 2007; Dorward et al. 2008; Snapp et al. 2010; Ayuke et al. 2011). There is evidence from Gambia, Madagascar, and Sierra Leone that the mean yields of rice under conventional farming are much lower than those under agroecology (Uphoff 2003, 2013). In Gambia, for example, conventional rice yields were reported at 2.3 t/ha, while those under agroecology recorded 7.1 t/ha (Uphoff 2003, 2013). Madagascar, on its part, recorded conventional rice yields of 2.6 t/ha, while agroecology recorded 7.2 t/ha; in Sierra Leone, conventional rice yields recorded 2.3 t/ha, and agroecology recorded 5.3 t/ha (Figure 3) (Uphoff 2003, 2013). Between 1994 and 1999 in Madagascar, the mean yields were higher for conventional farms, which recorded 8.55 t/ha, while peasant or agroecology farms recorded 2.36 t/ha (Uphoff 2003). Based on these initial examples, it can be said that the yields under agroecology are generally higher in the context of the countries under consideration, except for Madagascar.

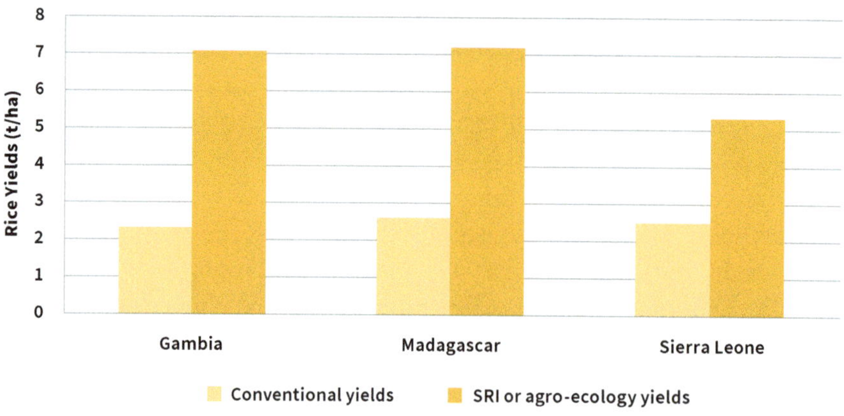

Figure 3. Mean rice yields under agroecology and conventional farming. Source: Authors' compilation based on data from Uphoff (2003, 2013).

These days, agriculture in Africa is being modernized, as seen in the increased use of improved varieties of planting materials, machines, pesticides, and inorganic fertilizers, among other aspects. It has been stated that the full capabilities of improved varieties are often much more effective when other external inputs, such as inorganic fertilizers, are employed (IFAD 1986). Therefore, most food security campaigns in Africa have emphasized the modernization of agriculture through inorganic inputs, and therefore conventional agriculture. It has been argued that, in Cameroon, agroecology nutrients are often not available for crops when compared to inorganic nutrients under conventional farming. Additionally, it has been argued that organic sources are unable to trigger an agricultural revolution in Africa because of the lack of valorization (IFAD 1986; Epule et al. 2012). It has been added that all attempts at increasing productivity in Cameroon without external inputs are likely to fail, because the optimal scenario for yield enhancement includes a scenario that combines both agroecology farming and conventional farming (FAO 1987). In most African countries, yield improvements are often based on increasing farm sizes. Agroecology will further reinforce this equation by taking up more land, while conventional agriculture drives intensive agriculture and therefore reduces the yield–land dilemma (Lindell et al. 2010a, 2010b; Jayne et al. 2003). However, conventional agricultural options, such as inorganic fertilizers, are more likely to pollute water resources (Lindell et al. 2010a, 2010b; Epule 2019).

For agroecology to be successful in Africa, its components need to be adequately valorized. For the twin challenges of improving yields and protecting the environment to be achieved through agroecology, all of the processes involving the enhancement of organic manure, compost, the use of prey–predator relationships to control pests, and the dependence on natural cycles must be valorized to levels parallel to those of conventional agriculture. With a few exceptions to the leading role of agroecology-related yields in Africa presented above, most reports on the success of agriculture in Africa are often linked to the role of the conventional options of agriculture. Africans need to be educated on how to valorize agroecology, and pilot agroecology projects need to be established in various countries. Countries such as Malawi, with some of the well-known agroecology projects, still indulge in the intensification of agriculture through external inputs in bids aimed at enhancing food security (Borlaug 2000; Denning et al. 2009). This is seen as Malawi has addressed food insecurity by enhancing investments in N fertilizers and high-yielding varieties, as well as by enhancing access to these (Dorward et al. 2008; Mäder et al. 2002; Gregory et al. 2005). It has been argued that the moderate application of fertilizer, of about 35 kg N ha^{-1}, doubled the grain production of an initially unfertilized farm: from 1.05 Mg ha^{-1} to about 2.17 Mg ha^{-1} (Snapp et al. 2010). Even in the Songani and Ekwendeni research sites in Malawi, agroecology-based yields of maize increased after the introduction of conventional inorganic fertilizers (Snapp et al. 2010). Much of this increase was attributed to external inputs. More evidence from Malawi shows that agroecology can only meet the twin challenges of feeding an increasing population and protecting the environment if aspects of access and scale are adequately valorized (Gregory et al. 2005). While there is enough information on conventional agricultural inputs, such as organic fertilizers, there is little information on the use of agroecology-related organic fertilizers. On the other hand, the rate of inorganic fertilizer use in Africa is generally lower when compared to the developed world due to issues related to costs, the absence of sufficient credit facilities, and unfavorable policies (Jayne et al. 2003). In Kabate, Central Kenya, reports hold that agroecology-related inputs are unable to sustain crop yields as well as the environment at their current level of valorization (Ayuke et al. 2011; Borlaug 2000; Denning et al. 2009; Chivenge et al. 2009).

New research is showing that the most feasible scenario of improving the crop yield in Africa now is through a combination of agroecology and conventional agricultural options (Figure 4) (Ayuke et al. 2011; Hafidi et al. 2012). This is because agroecology alone cannot enhance production to meet the food needs in Africa based on the current population growth and increased vulnerability to climate change. Agroecology, however, remains critical for small-scale farmers in the Global South due to their inability to secure large quantities of inorganic fertilizers (Heisey and Mwangi 1996; Demelash et al. 2014).

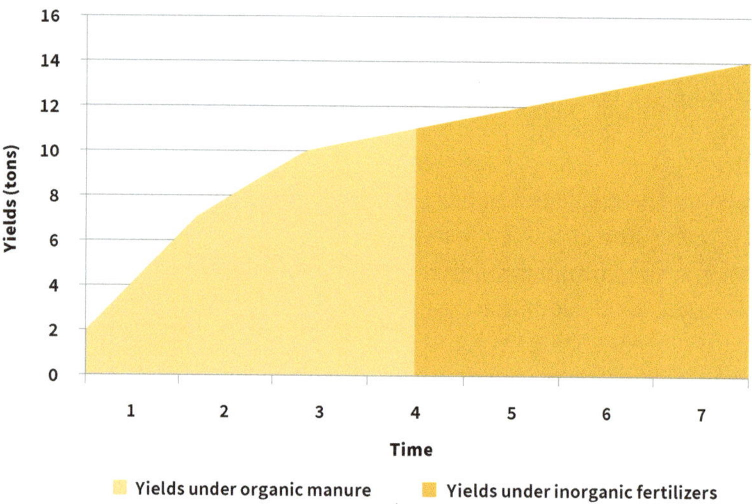

Figure 4. Evolution of yields under agroecology and conventional agriculture in Africa. Source: Authors' compilation based on data from Epule et al. (2015).

Though the best combination for yield improvements in Africa now involves a combination of agroecology and conventional agricultural practices, it is still important to report here that the increase is often minimal. A study carried out in Central Kenya shows that when agroecological inputs are added to maize farms, the yields increase and flatten out at some point, even if more inputs are added, whereas the introduction of conventional inputs, such as nitrogen fertilizers, was able to trigger the maize yield curve to rise further, though marginally (Gregory et al. 2005;

Heisey and Mwangi 1996; Demelash et al. 2014; Pereira et al. 2018). In Africa, most farming systems that do not employ the use of agroecology or conventional farming options usually experience low yields. Invariably, the optimal scenario for yield improvements is a combination of both agroecology and conventional inputs (Bationo and Mokwunye 1991; Bado et al. 1997, 2007). In Burkina Faso, for example, inputs of conventional fertilizers, such as nitrogen, calcium, and potassium (NKP), combined with agroecology-based options, such as manure, triggered higher yields during the years 2000, 2001, and 2003 (Figure 5). The alternatives show that in a scenario with only the use of NKP, the resultant yields were intermediate during the same period, while the "control" scenario with neither agroecology or conventional inputs showed high grain yields in the beginning because the soils had been left to fallow, and in later years showed a decline in yields (Figure 5). Similar results have been reported in Ethiopia (Demelash et al. 2014).

Agroecology needs to be valorized in Africa because of its enormous potentials. Compost, for example, has positive impacts on the physicochemical and biological properties of soils, which in turn help in driving improved soil quality and yields (Demelash et al. 2014). Unfortunately, as seen in the cases presented here, agriculture in Africa will continue to be driven by conventional inputs if agroecology is not valorized and issues of access to resources are not properly handled (Snapp et al. 2010; Rosegrant and Cline 2003; Razanakoto et al. 2021). Despite the prospects of agroecology in Africa, the system still has a lot of challenges. Since agriculture in Africa is mostly in the hands of smallholders, production is generally under natural conditions driven by limited access to conventional production inputs (Razanakoto et al. 2021). The valorization of agroecology to levels that are parallel to the status of conventional agriculture mandates a synergy between all agricultural stakeholders. The question should no longer be "can agroecology feed Africa's population", but instead be one focused on "how can agroecology be valorized to increase crop yields and protect the environment?" Evidently, there seems to be no easy way out of this agroecology–conventional agriculture dilemma. Perhaps the fact that these systems are driven by several drivers working together in the context of Africa makes this puzzle even more difficult (Mugwanya 2019). In the final part of this chapter, an examination of the benefits, limits, and challenges of these two broad paradigms is examined.

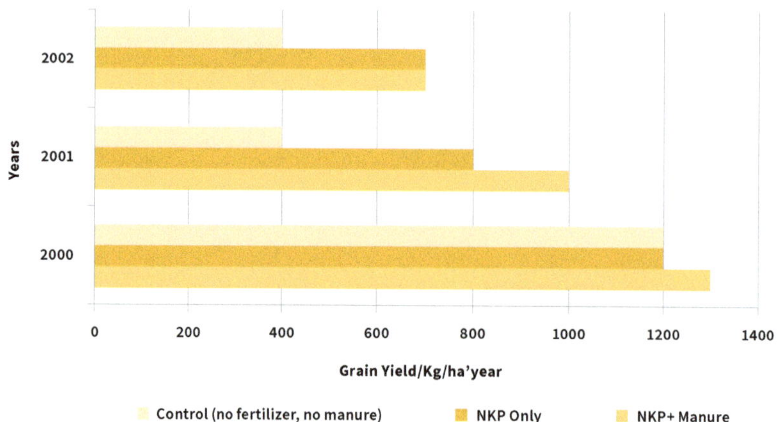

Figure 5. Evolution of grain yields under three scenarios in the Guinean zone of Burkina Faso. Source: Authors' compilation based on data from Bado et al. (2007); Demelash et al. (2014).

Therefore, as can be seen above, agroecology is good for the environment, but its current level of valorization does not make it able to overcome the food security crises that the continent is witnessing. For agroecology to fill the twin gaps of food security and environmental resilience, it should be valorized, as seen in the successful case studies presented above. On the other hand, conventional farming currently holds more prospects of reducing food insecurity, because under this approach greater yields are currently being obtained. However, the paradox is that despite the food security benefits that this approach currently offers, it is flawed due to its daunting effects on the environment. Consequently, the future of African agriculture rest in the valorization of agroecology to levels at which it can sustain actual and future food production needs, while at the same time reducing environmental degradation to a minimum.

6. Conclusions

This chapter has shown that agroecology has the potential of being more sustainable for the African environment when compared to conventional agriculture. However, the status of agroecology in Africa is one that needs valorization to ensure that its yields can be parallel to those of conventional agriculture. Even though agroecology has minimal effects on the environment, its overdependence on land expansion is likely to expose the environment to more deforestation and its accompanying effects. In fact, there are results that show that, in Africa, smallholder farmers in West, Central, and East Africa depend mostly on expanding their farm sizes to improve yields (Epule et al. 2022). Conventional agriculture, on the other hand, currently produces higher yields in the current African context, while unfortunately being less suitable for the environment. Therefore, the goal at this juncture will be to valorize agroecology to a level where it can sustain the twin benefits of ensuring food security while also sustaining the environment. In Africa, since most agricultural production is in the hands of smallholder farmers, the current access to conventional agricultural options is highly limited, thus affecting food security. Most of these smallholders cultivate without any major inputs and without well-valorized knowledge of agroecology. The bigger agro-industrial organizations are the main users of major conventional agricultural options, since these organizations can afford such conventional inputs. Such organizations would also benefit from the advantages of the valorization of agroecology. In the context of future research, new ways must be researched and developed on how to make conventional agriculture cleaner and more sustainable, while pilot farms that will drive the valorization of the components of agroecology should also be accentuated. This implies that all stakeholders, including smallholder farmers, should be included in the design, elaboration, execution, evaluation, and monitoring of such efforts in attempts at creating ownership, which is a key element of success and acceptance. Therefore, a policy approach that is both top-down and, most importantly, bottom-up will go a long way to ensure success. The main weakness of this work is that it mainly uses a review approach that uses both peer-reviewed and grey literature. Grey literature is non-standardized, and therefore less reliable and acceptable. As a result, it is important for more field-based studies to be conducted to investigate these initial literature-driven results at different scales across Africa.

Author Contributions: Conceptualization, T.E.E.; Methodology, T.E.E.; Software, T.E.E.; Formal Analysis, T.E.E. and A.C.; Investigation, T.E.E. and A.C.; Resources, AC.; Data Curation, T.E.E. and A.C.; Writing—Original Draft Preparation, T.E.E.; Writing—Review and Editing, A.C.; Supervision, A.C. and T.E.E.; Project Administration, A.C. All authors have read and agreed to the published version of the manuscript.

Funding: This work was funded by the Pan Moroccan Yield and Precipitation Gaps Project (PAMCPP) awarded to T.E.E. in 2021 at Mohammed VI Polytechnic University.

Acknowledgments: This work was made possible thanks to An APRA Grant to the PAMOCPP Project under grant number 149DPRA01.

Conflicts of Interest: The authors declare no conflict of interest.

References

Abedi, Tayebeh, Abass Alemzadeh, and Seyed Abdolreza Kazemeini. 2010. Effect of organic and inorganic fertilizers on grain yield and protein banding pattern of wheat. *Australian Journal of Crop Science* 4: 384–89.

Altieri, Miguel A. 1995. *Agroecology*, 2nd ed. Boulder: Westview Press.

Altieri, Miguel A. 2002. Agroecology: The science of natural resource management for poor farmers in marginal environments. *Agriculture, Ecosystems & Environment* 93: 1–24.

Altieri, Miguel A. 2009. Agroecology, small farms, and food sovereignty. *Monthly Review* 61: 102–13. [CrossRef]

Ayuke, Frederick Ouma, Lijbert Brussaard, Benard Vanlauwe, Johan Six, David Kiprono Lelei, Catherine Kibunja, and Mirjam Pulleman. 2011. Soil fertility management: Impacts on soil macrofauna, soil aggregation and soil organic matter allocation. *Applied Soil Ecology* 48: 53–62. [CrossRef]

Badgley, Catherine, Jeremy Moghtader, Eileen Quintero, Emily Zakem, Jahi Chappell, Katia Aviles-Vazquez, Andrea Samulon, and Ivette Perfecto. 2007. Organic agriculture and the global food supply. *Renewable Agriculture and Food Systems* 22: 86–108. [CrossRef]

Bado, Boubie Vincent, André Bationo, Francois Lompo, Michel Cescas, and Michael P. Sedogo. 2007. Mineral fertilizers, organic amendments, and crop rotation managements for soil fertility maintenance in the Guinean zone of Burkina Faso (West Africa). In *Advances in Integrated Soil Fertility Management in Sub-Saharan Africa: Challenges and Opportunities*. Dordrecht: Springer, pp. 171–77.

Bado, Boubie Vincent, Michel Papdoba Sédogo, Michel Pierre Cescas, François Lompo, and André Bationo. 1997. Effet à long terme des fumures sur le sol et les rendements du maïs au Burkina Faso. *Cahiers Agricultures* 6: 571–75.

Bationo, André, and Uzo Mokwunye. 1991. Role of manures and crop residue in alleviating soil fertility constraints to crop production: With special reference to the Sahelian and Sudanian zones of West Africa. *Fertilizer Research* 29: 117–25. [CrossRef]

Borlaug, Norman E. 2000. Ending world hunger. The promise of biotechnology and the threat of antiscience zealotry. *Plant Physiology* 124: 487–90. [CrossRef]

Bumb, Balu L., and Rob Groot. 2004. Improving fertilizer supply in sub-Saharan Africa. Paper presented at US–Africa Agribusiness Conference, Monterey, CA, USA, 13–16 November; pp. 7–10.

Chivenge, Pauline, Bernard Vanlauwe, Roberta Gentile, Hellen Kamiri Wangechi, Daniel Njiru Mugendi, Van Kessel-Bakvis, and Johan Six. 2009. Organic and mineral input management to enhance crop productivity in Central Kenya. *Agronomy Journal* 101: 1266–75. [CrossRef]

Demelash, Nigus, Wondimu Bayu, Sitot Tesfaye, Feras Ziadat, and Rolf Sommer. 2014. Current and residual effects of compost and inorganic fertilizer on wheat and soil chemical properties. *Nutrient Cycling in Agroecosystems* 100: 357–67. [CrossRef]

Denning, Glenn, Patrick Kabambe, Pedro Sanchez, Alia Malik, Rafael Flor, Rebbie Harawa, Phelire Nkhoma, Colleen Zamba, Clement Banda, Chrispin Magombo, and et al. 2009. Input subsidies to improve smallholder maize productivity in Malawi: Toward an African green revolution. *PLoS Biology* 7: e1000023. [CrossRef] [PubMed]

Dorward, Andrew, Ephraim Chirwa, Valerie A. Kelly, Thomas S. Jayne, Rachel Slater, and Duncan Boughton. 2008. *Evaluation of the 2006 Agricultural Inputs Subsidy Programme, Malawi: Final Report*. London: School of Oriental and African Studies.

Epule, Epule Terence, Changhui Peng, Laurent Lepage, Balgah Sounders Nguh, and Ndiva Mongoh Mafany. 2012. Can the African food supply model learn from the Asian food supply model? Quantification with statistical methods. *Environment, Development and Sustainability* 14: 593–610. [CrossRef]

Epule, Epule Terence, Christopher Robin Bryant, Cherine Akkari, and Oumarou Daouda. 2015. Can organic fertilizers set the pace for a greener arable agricultural revolution in Africa? Analysis, synthesis, and way forward. *Land Use Policy* 47: 179–87. [CrossRef]

Epule, Terence Epule. 2019. Contribution of Organic Farming Towards Global Food Security: An Overview. *Organic Farming* 1: 1–16.

Epule, Terence Epule, Abdelghani Chehbouni, and Driss Dhiba. 2022. Recent Patterns in Maize Yield and Harvest Area across Africa. *Agronomy* 12: 374. [CrossRef]

Food and Agriculture Organization (FAO) of the United Nations. 1987. *Agriculture: Toward 2000*. Economic and Social Development Series; Rome: Food and Agriculture Organization (FAO) of the United Nations.

Gliessman, Stephen R., Eric Engles, and Robin Krieger. 1998. *Agroecology: Ecological Processes in Sustainable Agriculture*. Boca Raton: CRC Press.

Gregory, Peter J., John S. I. Ingram, and Michael Brklacich. 2005. Climate change and food security. *Philosophical Transactions of the Royal Society B: Biological Sciences* 360: 2139–48. [CrossRef]

Hafidi, Mohamed, Soumia Amir, Abdelilah Meddich, Abdelmajid Jouraiphy, Peter Winterton, Mohamed El Gharous, and Robin Duponnois. 2012. Impact of applying composted biosolids on wheat growth and yield parameters on a calcimagnesic soil in a semi-arid region. *African Journal of Biotechnology* 11: 9805–815.

Heisey, Paul, and William Mwangi. 1996. *Fertilizer Use and Maize Production in Sub-Saharan Africa*. Mexico City: CIMMYT.

Hossain, Singh, and Ved Prakash Singh. 2000. Fertilizer use in Asian agriculture: Implications for sustaining food security and the environment. *Nutrient Cycling in Agroecosystems* 57: 155–69. [CrossRef]

International Fund for Agricultural Development (IFAD). 1986. *Cameroon Fertilizer Sector Study*. Muscle Shoals: International Fertilizer Development Center, May.

Jayne, Thomas S., Valerie A. Kelly, and Eric W. Crawford. 2003. *Fertilizer Consumption Trends in Sub-Saharan Africa*. No. 1095-2016-88171. East Lansing: Department of Agricultural, Food, and Resource Economics, Michigan State University.

Kerr, Rachel Bezner, Sieglinde Snapp, Marko Chirwa, Lizzie Shumba, and Rodgers Msachi. 2007. Participatory research on legume diversification with Malawian smallholder farmers for improved human nutrition and soil fertility. *Experimental Agriculture* 43: 437–53. [CrossRef]

Lindell, Lina, Mats Åström, and Tomas Öberg. 2010a. Land-use change versus natural controls on stream water chemistry in the Subandean Amazon, Peru. *Applied Geochemistry* 25: 485–95. [CrossRef]

Lindell, Lina, Mats E. Åström, and Sirkku Sarenbo. 2010b. Effects of forest slash and burn on the distribution of trace elements in floodplain sediments and mountain soils of the Subandean Amazon, Peru. *Applied Geochemistry* 25: 1097–106. [CrossRef]

Mäder, Paul, Andreas Fliessbach, David Dubois, Lucie Gunst, Padruot Fried, and Urs Niggli. 2002. Soil fertility and biodiversity in organic farming. *Science* 296: 1694–97. [CrossRef]

Matson, Pamela A., William J. Parton, Allison G. Power, and Mike J. Swift. 1997. Agricultural intensification and ecosystem properties. *Science* 277: 504–9. [CrossRef]

Morris, Michael, Valerie A. Kelly, Ron J. Kopicki, and Derek Byerlee. 2007. *Fertilizer Use in African Agriculture: Lessons Learned and Good Practice Guidelines*. Directions in Development; Agriculture and Rural Development. Washington, DC: World Bank. Available online: https://openknowledge.worldbank.org/handle/10986/6650 (accessed on 12 August 2021).

Mugwanya, Nassib. 2019. Why agroecology is a dead end for Africa. *Outlook on Agriculture* 48: 113–16. [CrossRef]

Pereira, Laura, Rachel Wynberg, and Yuna Reis. 2018. Agroecology: The future of sustainable farming? *Environment: Science and Policy for Sustainable Development* 60: 4–17. [CrossRef]

Pichot, Jean-Pascal, Michel Papaoba Sedogo, Jean-François Poulain, and Jacques Arrivets. 1981. Evolution de la fertilité d'un sol ferrugineux tropical sous l'influence de fumures minérales et organiques. *L'agronomie Tropicale* 36: 122–33.

Pretty, Jules. 1995. *Regenerating Agriculture*. London, Earthscan and Washington, DC: National Academy Press.

Razanakoto, Onjaherilanto R., Sitrakiniaina Raharimalala, Eddy J. Randriamihary Fetra Sarobidy, Jean-Chrysostome Rakotondravelo, Patrice Autfray, and Hanitriniaina M. Razafimahatratra. 2021. Why smallholder farms' practices are already agroecological despite conventional agriculture applied on market-gardening. *Outlook on Agriculture* 50: 80–89. [CrossRef]

Reid, John F., John Schueller, and William R. Norris. 2003. Reducing the Manufacturing and Management Costs of Tractors and Agricultural Equipment. *Agricultural Engineering International: CIGR Journal* 5: 1–12.

Rosegrant, Mark W., and Mark Svendsen. 1993. Asian food production in the 1990s: Irrigation investment and management policy. *Food Policy* 18: 13–32. [CrossRef]
Rosegrant, Mark W., and Sarah A. Cline. 2003. Global food security: Challenges and policies. *Science* 302: 1917–19. [CrossRef]
Seufert, Verena, Navin Ramankutty, and Jonathan A. Foley. 2012. Comparing the yields of organic and conventional agriculture. *Nature* 485: 229–32. [CrossRef]
Snapp, Sieglinde S., Malcolm J. Blackie, Robert A. Gilbert, Rachel Bezner-Kerr, and George Y. Kanyama-Phiri. 2010. Biodiversity can support a greener revolution in Africa. *Proceedings of the National Academy of Sciences* 107: 20840–45. [CrossRef]
Stoop, Willem A., Norman Uphoff, and Amir Kassam. 2002. A review of agricultural research issues raised by the system of rice intensification (SRI) from Madagascar: Opportunities for improving farming systems for resource-poor farmers. *Agricultural Systems* 71: 249–74. [CrossRef]
Tilman, David, Kenneth G. Cassman, Pamela A. Matson, Rosamond Naylor, and Stephen Polasky. 2002. Agricultural sustainability and intensive production practices. *Nature* 418: 671–77. [CrossRef]
Trewavas, Anthony. 2001. Urban myths of organic farming. *Nature* 410: 409–10. [CrossRef]
Uphoff, Norman. 2003. Higher yields with fewer external inputs? The system of rice intensification and potential contributions to agricultural sustainability. *International Journal of Agricultural Sustainability* 1: 38–50. [CrossRef]
Uphoff, Norman, ed. 2013. *Agroecological Innovations: Increasing Food Production with Participatory Development*. London: Routledge.
Valenzuela, Hector. 2016. Agroecology: A global paradigm to challenge mainstream industrial agriculture. *Horticulturae* 2: 2. [CrossRef]

identifying desired functions of the farming system; (4) what the resilience capacities are, by assessing resilience capacities; and (5) what enhances resilience, by assessing resilience-enhancing attributes (Meuwissen et al. 2019).

The objective of this review is to provide an overview and evaluate the role of agroforestry in enhancing agricultural resilience and productivity. Specifically, the review aims to carry out the following:

1. Examine the scientific basis and principles underlying agroforestry as a sustainable farming practice;
2. Investigate the practical applications of agroforestry in diverse farming systems, including agri-silvicultural, silvo-pastoral, and agro-silvi-pastoral systems;
3. Explore the impact of agroforestry on livelihood improvement;
4. Identify potential challenges and opportunities in scaling up agroforestry for widespread adoption and its implications for achieving sustainable and resilient farming systems.

2. Different Types of Agroforestry Systems

Considering the spatial and temporal arrangement of its components, agroforestry systems can be categorized into three main systems: agri-silvicultural systems, silvo-pastoral systems, and agro-silvi-pastoral systems (Nair et al. 2021).

Agri-silvi-cultural systems involve the intentional integration of crops and trees within the same farming landscape (Figure 1a) (Nair et al. 2021). The trees can be fruit trees, timber trees, or other commercially viable tree species. The agricultural component may include food crops, cash crops, medicinal plants, and mushrooms and is popular as a low-input production system (Atangana et al. 2014a). The integration of agriculture and forestry components allows for complementary interactions and synergies. Intercropping, alley cropping, shifting cultivation, chena, taungya, home gardening, plantation crop combinations, trees for conservation, shelter belts, windbreak, and live hedges can be identified as examples of this cropping system (Ayyam et al. 2019).

The silvo-pastoral systems are an integrated form of land use that combines trees, forage/pasture crops, and/or livestock grazing (Figure 1b) (Nair et al. 2021). They involve the deliberate and simultaneous management of trees and pasture for the production of food, timber, forage, and livestock. Fodder production is the main productive function in this system and needs a comparatively higher involvement of socio-economic and management than agri-silvicultural systems do (Atangana et al. 2014a). Silvo-pastoral systems can provide multiple benefits, including improved animal welfare, increased productivity, enhanced environmental sustainability, and economic advantages. Silvi-pastures, horti-pastoral patures, trees on rangelands, plantation crops with pastures, protein banks, and seasonal forestry grazing can be categorized under this system (Ayyam et al. 2019; Dhyani et al. 2021).

Agro-silvi-pastoral systems integrate crops, trees, and livestock within the same farming landscape (Figure 1c) (Nair et al. 2021). The agricultural component may include food crops, cash crops, or forage crops for livestock. The tree component may include multipurpose trees. As a livestock component, cattle, buffaloes, swine, goat, sheep, and poultry farming are incorporated into the system. The integration of these components allows for complementary interactions and synergies between agriculture, silviculture, and livestock production. This system requires high human involvement for success.

Furthermore, apiculture with trees (integrating trees with bee-keeping) (Figure 1d) and aqua forestry (integrating trees with fish) are also common in some areas of the world, especially in highland sub-humid tropical areas (Ayyam et al. 2019). Agri-silvi-cultural systems are primarily suited for lowland humid tropical environments, while silvo-pastoral and agro-silvi-pastoral systems are more apt for highland humid tropics above 1200 m from sea level and lowland sub-humid tropical climates, respectively (Atangana et al. 2014a). Agri-silvi-pastoral, agro-pastoral, and silvo-pastoral systems are mostly prevalent in parts of Africa, Asia, and the Americas, whereas some gardens are widely practiced in South and Southeast Asia (Mahmud et al. 2021). In temperate zones, silvo-pastoral systems with coniferous trees and livestock are the most prevalent type of agroforestry system (Den Herder et al. 2017). Other than that, native agroforestry systems such as forests and woodlots and mesquite-based systems in North america, nomadic systems in South America, silvi-arable systems in temperate zones in China, and Dehesa systems in europe are some of the common agroforestry systems which can be seen in temperate zones in the world except shelter belts, windbreaks, riparian forest systems, intercropping, and allycropping systems (Bhardwaj et al. 2017).

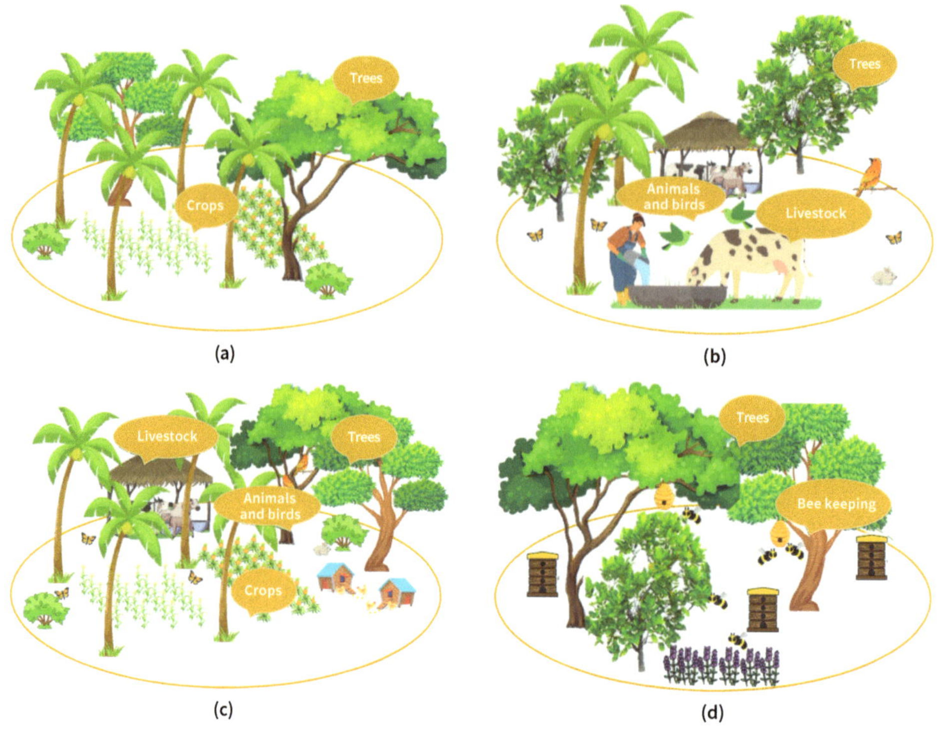

Figure 1. Different agroforestry systems: (**a**) agri-silvicultural systems; (**b**) silvo-pastoral system; (**c**) agro-silvi-pastoral systems; (**d**) apiculture. Source: Figure by authors.

3. Potential Benefits from Agroforestry Farming

3.1. Biodiversity and Ecosystem Services

Agroforestry systems play a vital role in promoting biodiversity within agricultural landscapes, leading to enhanced resilience and productivity (Octavia et al. 2022; Udawatta et al. 2019). High biodiversity with the presence of diverse plant species, including trees, crops, and understory vegetation, creates a complex habitat that supports a wide range of plant and animal species (Rahman et al. 2023). The presence of diverse plant species in agroforestry systems attracts a greater variety of beneficial insects, birds, and other predators that act as natural enemies of pests (Rahman et al. 2023; Suroso et al. 2023). These beneficial organisms contribute to pest control by preying on or parasitizing pests, reducing the need for synthetic pesticides (Fahad et al. 2022; Jose 2009). For example, trees in agroforestry systems can provide shelter and food sources for predatory insects, such as ladybugs, lacewings, and parasitic wasps, which help control crop pests. Agroforestry systems, with their various plant communities, support a broader range of pollinators, including bees, butterflies, birds, and bats (Udawatta et al. 2019). These pollinators play a crucial role in the reproductive success of both crops and wild plant species, contributing to higher yields and improved fruit sets (Bentrup et al. 2019; Centeno-Alvarado et al. 2023; Klein et al. 2003). The presence of flowering trees and shrubs in agroforestry systems provides additional nectar and pollen sources for pollinators, increasing their abundance and diversity (Lee-Mader et al. 2020). Trees in agroforestry systems contribute to an excellent biodiversity corridor by providing interconnected and ecologically diverse habitats that facilitate wildlife movement between fragmented landscapes (Haggar et al. 2019). As an overall effect, this system increases food security and food diversification.

3.2. Nutrient Cycling and Soil Health

Agroforestry systems improve nutrient cycling and soil health through various mechanisms (Atapattu et al. 2017; Dissanayaka et al. 2023). The leaf litter from trees, along with the decomposition of organic matter from different plant species, contributes to increased soil organic matter content. This organic matter enhances soil structure, water holding capacity, nutrient availability, and microbial activity (Mohammadi et al. 2011). The diverse root systems of trees and crops in agroforestry systems facilitate nutrient uptake from different soil depths, reducing nutrient leaching and enhancing nutrient cycling within the ecosystem (Hugenschmidt and Kay 2023). Furthermore, the diverse root systems of trees, along with the associated mycorrhizal networks, where tree roots form symbiotic relationships with beneficial fungi, increase nutrient availability for surrounding crops and improve crop productivity. They interact with the soil biological and chemical properties to extract and redistribute water in ways that remove pollutants from the contaminated soils over time through methods like decontamination, hyperaccumulation, and hydraulic lift (Table 1) (Atangana et al. 2014b; François et al. 2023). Root exudates from trees contain organic compounds that fuel microbial populations in the rhizosphere, promoting nutrient transformations and facilitating nutrient availability to plants (Beule et al. 2022).

Table 1. The ability of various agroforestry systems to reduce soil pollution levels.

Category of Agroforestry System	Pollution Abatement-Related Main Observation	Reference
Intercropping systems with *Acacia ampliceps* or *Azadiracta indica*	Reduction in Pb contamination in municipal wastewater and industrial wastewater	Hussain et al. (2020)
Tree alley cropping systems	Pesticides showed disappearance rates of more than 61.5% and up to 100% in the points closest to the tree row	Pavlidis et al. (2020)
Tree alley cropping systems	Reduction in nitrate pollution in groundwater	Gikas et al. (2016)
Tree-based agroforestry systems	Reduction in nitrogen and phosphorus residues in soil by 20% and up to 100%, and pesticide transport up to 100%	Pavlidis and Vassilios (2017)
Tree-based agroforestry systems	Purifying the atmosphere by reducing net emissions of nitrous oxide (by drying the soil, increasing soil aeration, and removing excess nitrate concentrations) and ammonia gases (by reducing emissions and recapturing through tree foliage)	Lawson et al. (2020)

Source: Table by authors.

The presence of trees and diverse vegetation in agroforestry systems improves water infiltration rates and helps control soil erosion (Fahad et al. 2022). The tree canopy intercepts rainfall, reducing the impact of raindrops on the soil surface and minimizing erosion. The deep roots of trees contribute to improved soil structure and porosity, enhancing water infiltration and reducing surface runoff (Jinger et al. 2022). Agroforestry systems thus mitigate the risk of soil erosion, especially during heavy rainfall events, helping to maintain soil fertility and prevent nutrient loss.

Nitrogen-fixing plants play a significant role in agroforestry systems by enriching the soil with nitrogen, a vital nutrient for plant growth (Sarvade and Singh 2014; Thomas et al. 2018). These plants have the unique ability to convert atmospheric nitrogen gas into a usable form through a symbiotic relationship with nitrogen-fixing bacteria and are genetically determined (Table 2) (Suzaki et al. 2015). These bacteria possess the enzyme nitrogenase, which enables them to convert atmospheric nitrogen (N_2) into ammonia (NH_3) through a process known as nitrogen fixation (Hoffman et al. 2014). The ammonia is then converted into forms such as ammonium (NH_4^+) that can be utilized by plants. As a result, the presence of nitrogen-fixing plants improves the overall nitrogen availability in the soil, supporting the growth and productivity of neighboring plants.

Table 2. Nitrogen-fixing plant species and its nitrogen-fixing potential.

Scientific Name	The Woodiness of the Plant	Nitrogen Fixed (kg N/ha/year)
Acacia albida	Woody	20
Acacia mearnsii	Woody	60–110
Calliandra calothyrsus	Woody	24–93
Erythrina poeppigiana (Walp.)	Woody	60
Erytriana fusca	Woody	80
Gliricidia sepium	Woody	13
Leucaena leucocephala	Woody	100–500
Sesbania sesban	Woody	18–8
Sesbania rostrata	Woody	70–458
Glycine max L. Merr	Herbaceous	60–168
Medicago sativa L.	Herbaceous	229–290
Trifolium spp.	Herbaceous	128–207
Vigna sinensis L. (Walp.)	Herbaceous	73–354

Source: Authors' compilation based on data from Adams et al. (2010); Muoni et al. (2020); Nuwarapaksha et al. (2023); Sarvade et al. (2014); Sharma et al. (2022).

3.3. Livelihoods and Economic Benefits

Agroforestry systems provide a wide range of products that can be harvested and sold, diversifying income streams for farmers. These products include timber from trees, fruits and nuts from fruit trees, fodder for livestock, livestock products (ex: meat, milk, and wool), and various non-timber forest products such as medicinal plants, honey, and bamboo (Kuyah et al. 2020; Tsegaye 2023). The diversity of products allows farmers to access multiple markets and income sources, reducing their dependence on a single crop or commodity. Agroforestry has demonstrated its potential for poverty reduction and creating sustainable economic opportunities, particularly in rural areas. By diversifying income sources and providing multiple products, agroforestry systems can increase the resilience of farming communities and reduce their vulnerability to market fluctuations. Agroforestry also offers long-term benefits by enhancing soil fertility, reducing erosion, and promoting sustainable land management practices, ensuring the productivity and economic viability of farming systems over time (Dissanayaka et al. 2023).

3.4. Climate Resilience and Adaptation

Trees in agroforestry systems provide shade and act as windbreaks, offering protection to crops and livestock (Santoro et al. 2020; Smith et al. 2021). During heatwaves or periods of high temperatures, the shade provided by trees can reduce heat stress in crops and animals, mitigating potential yield losses and negative impacts on livestock health (Ramil Brick et al. 2022). Furthermore, the presence of trees modifies the microclimate by creating more favorable conditions, such as reduced temperature fluctuations and increased humidity, within the agroforestry system (Rosenstock et al. 2019). For example, the jackfruit-based agroforestry system reported a 3.37–9.25% reduction in soil temperature in Bangladesh (Riyadh et al. 2018). Agroforestry systems contribute to climate change adaptation by sequestering carbon dioxide (CO_2) from the atmosphere, thereby mitigating greenhouse gas emissions (Chavan et al. 2021). Trees can capture and store carbon in their biomass and soil, acting as "carbon sinks". Differences in growth rates, wood densities, and lifespans are key factors influencing a tree's carbon storage capacity (Sharma et al. 2016). Certain tree species exhibit rapid growth, accumulating biomass swiftly, whereas others grow slowly but boast denser wood, leading to greater long-term carbon storage (Kaul et al. 2010). Agroforestry systems, with their tree components, can sequester significant amounts of carbon over time, thus helping to offset emissions and mitigate climate change (Table 3).

Table 3. Carbon storage potential of different agroforestry systems in different regions of the world.

Type of Agroforestry System	Established Region	Carbon Storage Potetial (Mg C ha^{-1}) *
Agro-silvicultural	Africa	29–53
	South America	39–102
		39–195
	Southeast Asia	12–228
		68–81
Silvi-pastoral	Australia	28–51
	North America	133–154
		104–198
	Northern Asia	15–18

* Carbon storage values were standardized to a 50-year rotation. Source: Authors' compilation based on data from Albrecht and Kandji (2003).

4. Factors to Consider

The implementation of agroforestry involves several key steps to ensure the successful establishment and management of the system. While the specific steps may vary depending on the context and objectives, the following considerations are the main steps commonly involved in agroforestry implementation.

4.1. Site Selection and Planning

The first step is to identify suitable sites for agroforestry implementation. Site assessment helps determine the compatibility of tree species, crops, and livestock in a specific area. Planning involves determining the appropriate design, layout, and arrangement of the agroforestry system to maximize benefits and minimize potential conflicts. The design and layout should be adapted to the agroecological region and should be easy to manage. Chuma et al. (2021) and Nath et al. (2021) found the importance of basic parameters including climate, elevation, soil, aspects, slope orientation, euclidean distance to the road, euclidean distance to a river, and euclidean distance to the town while implementing agroforestry systems. Among them, sunlight availability, the direction of wind, precipitation frequency and amount, and irrigation water availability should be prioritized (De Zoysa 2022).

4.2. Component Selection

Selecting suitable animal, tree, and crop species is crucial for the success of the agroforestry system. The selection should consider local conditions, market demand, ecological requirements, and compatibility with other components (Banyal et al. 2018). It is important to choose tree and livestock species that are well adapted to the climate, water, and light contents, have economic value, and provide ecosystem services (Nerlich et al. 2013). Those species should be compatible with the tree species and meet the specific objectives of the system (De Zoysa 2022).

4.3. Planting and Establishment

The next step involves planting and establishing the trees, crops, and any additional components of the system, such as forage or livestock. Proper planting techniques, including spacing, depth, and care, are essential for successful establishment. Adequate soil preparation, irrigation, and weed control measures should be implemented to promote plant growth and survival (Ngarava et al. 2022).

4.4. Management and Maintenance

Ongoing management and maintenance are crucial to increasing the health and productivity of the agroforestry system (Isaac et al. 2007). This includes practices such as pruning, thinning, pest and disease management, irrigation, and fertilization (Ngarava et al. 2022). Regular monitoring of the system helps identify any issues or adjustments needed to optimize performance and minimize potential conflicts between tree and crop components.

4.5. Harvesting and Utilization

The harvesting of tree products, such as timber, fruits, nuts, or non-timber forest products, and the utilization of crops and livestock occur at appropriate stages of growth and maturity. Proper harvesting techniques and post-harvest handling are important to maintain the quality and value of the products (Facheux et al. 2007). The utilization of harvested products may involve marketing, processing, or value addition activities.

5. Challenges with Agroforestry

Agroforestry systems can take many years to become fully productive, as woody perennials like trees and shrubs take time to establish and grow. This long payback period is a major limitation, as farmers do not see returns on their investment for the first several years (Sagastuy and Krause 2019). During this establishment phase, agroforestry systems often produce less than conventional agricultural systems do, reducing yields and farm income in the short term (Sollen-Norrlin et al. 2020). Managing the complexity of agroforestry systems with their multiple components can be challenging, as can balancing the resource demands of all the components through competition for light, water, and nutrients (Sollen-Norrlin et al. 2020). Farmers require substantial knowledge and skills to successfully manage this complexity. However, technical assistance and guidance are often lacking, due to the limited research and extension support for agroforestry (Kiyani et al. 2017). The high diversity of plants in agroforestry systems can attract some pests, diseases, and weeds compared to those in simpler systems. The availability of alternate hosts and resources can allow pest populations to build up (Griffiths et al. 1998; Staton et al. 2022). Controlling these requires additional management efforts. There are also challenges in marketing the diverse mix of products from agroforestry systems compared to commodity crops. A lack of established supply chains and low consumer awareness limit market opportunities in many areas (Do et al. 2020). This marketing uncertainty increases the risks and limits the economic viability of agroforestry. Additional challenges like limited financial resources, labor, planting materials, extreme weather events, livestock damage, and fire risks can all undermine agroforestry success (Ajayi 2007). While agroforestry offers benefits, overcoming its limitations requires substantial knowledge, resources, and long-term commitment from farmers and supporting institutions. It is important to note that the significance of these challenges can vary substantially across different geographical regions and socio-economic contexts.

6. To Improve the System

This complex system can be scaled up for resilient and productive farming by expanding the adoption and implementation of agroforestry practices at larger scales.

6.1. Proper Awareness and Knowledge Sharing

Raising awareness about the benefits and potential of agroforestry is crucial (Buck et al. 2020; Musa et al. 2019). This involves disseminating information about successful agroforestry case studies, research findings, and best practices. For that, workshops, seminars, forum discussions, training, peer assistance or advice, film shows, and action reviews by experts, targeting farmers, agricultural communities, and local and foreign stakeholders can be organized with the help of universities, research stations, and other governmental and non-governmental organizations (Kommey and Fombad 2023). Agroforestry demonstration farms can be established in different regions. These farms can showcase various agroforestry models, providing tangible examples of how agroforestry works and its benefits. Similarly, open days where farmers and the general public can visit these model farms can be arranged. Interactive sessions, guided tours, and hands-on experiences can significantly enhance understanding. Online platforms and resources can be maintained where farmers and communities can access knowledge and training on agroforestry techniques, management strategies, and potential income streams. Regularly updating these types of platforms with the latest news is necessary. In addition to that, monitoring and evaluation with continuous data collection and feedback loops will help to assess the needs and challenges faced by farmers practicing agroforestry and trading agroforestry products.

6.2. Policy Support and Enabling Environment

Supportive policies and institutional frameworks are essential for scaling up agroforestry (Dagar et al. 2020; Ndlovu and Borrass 2021). Governments and policymakers can create incentives, regulations, and programs that promote and facilitate agroforestry adoption (Xue 2023). This can include financial incentives, technical assistance, land tenure security, and favorable market conditions for agroforestry products (Buck et al. 2020). Collaboration among different government departments, such as agriculture, forestry, and environment, is critical for integrated policy support.

Access to finance and resources is crucial for farmers to adopt and scale up agroforestry practices (Foster and Neufeldt 2014; Simelton et al. 2017). Financial mechanisms, such as subsidies, grants, microfinance, and investment schemes, can be established to support agroforestry initiatives (Sagastuy and Krause 2019). Farmers also need access to quality planting material, technical support, and training. According to a survey conducted by Kareem et al. (2017) to determine the issues with agroforestry farming, the biggest problems in implementing agroforestry systems are insufficient capital, insufficient rainfall, land accessibility, an insufficient market center, a high cost of agro-chemicals, inaccessibility of tractors, and an instability of market price. Strengthening extension services and establishing nurseries or seed banks for agroforestry species can facilitate access to resources (Shah 2023).

Other than that, scaling up agroforestry requires a landscape-level approach, considering the spatial integration of agroforestry systems within larger farming landscapes and collaboration among various stakeholders, including farmers, researchers, government agencies, NGOs, and private sector actors. Suggestions of the farmers in Musebeya sector, Nyamagabe District, in the southern province of Rwanda to improve the agroforestry systems are shown in Figure 2.

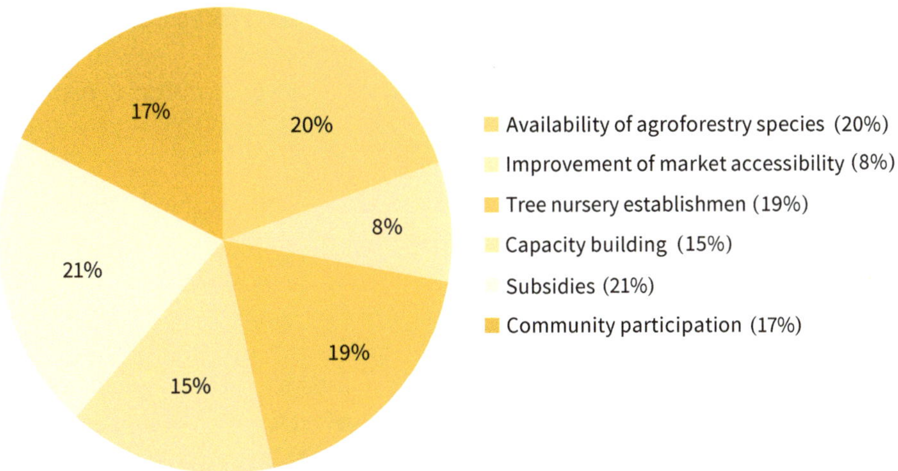

Figure 2. Suggestions of the farmers in Musebeya sector, Nyamagabe District, in the southern province of Rwanda to improve the agroforestry systems. Source: Authors' compilation based on data from Kiyani et al. (2017).

7. The Contribution of Agroforestry to Sustainable Development Goal 2

As a holistic and sustainable agricultural approach, agroforestry plays a vital role in advancing the objectives of SDG 2 and moving toward a world with zero hunger (Montagnini and Metzel 2017). Diversified and nutritious food production with integrated crop and livestock management and the minimum addition of synthetic harmful agro-inputs fulfill the basic requirement of SDG 2, ensuring access to safe, nutritious, and sufficient food for all people (Duffy et al. 2021). Improved soil fertility and land productivity via promoting natural fertilization processes like legume cropping and high organic matter addition in the soil indirectly influence higher yields, which contribute to increased food availability (Fahad et al. 2022). Agroforestry systems are often more resilient to climate change due to their biodiversity and ability to conserve water. In the face of changing climate patterns, agroforestry provides a sustainable way to ensure food production, aligning with the goals of SDG 2, enhancing adaptive capacity against climate-related hazards and natural disasters (Mbow et al. 2014; Platis et al. 2019). In addition to that, agroforestry encourages local food production. When food is produced locally, there is often less waste associated with transportation and storage, contributing to the aim of reducing food loss and waste. Importantly, agroforestry can provide additional sources of income through the sale of tree products, such as fruits, nuts, and timber (Beetz 2011; Kassie 2018). Increased income diversity for farmers improves their access to food and supports their ability to provide nutritious meals for their families. This sustainable land use system reduces soil erosion, conserves water, and protects biodiversity, contributing to the objective of SDG 2 on ensuring the sustainable use of terrestrial ecosystems.

8. Conclusions

In conclusion, resilient and productive farming is crucial to meeting the global challenge of ensuring food security for a growing population. Agroforestry practices play a vital role in achieving this objective by integrating trees, crops, and livestock in innovative and sustainable ways. Agroforestry systems can provide numerous benefits, including biodiversity conservation, nutrient cycling, soil health improvement, multiple income streams, climate resilience, and carbon sequestration. Through the integration of diverse components, such as agri-silvicultural, silvo-pastoral, and agro-silvi-pastoral systems, agroforestry can enhance agricultural productivity, support sustainable livelihoods, and contribute to ecological sustainability. Scaling up agroforestry requires raising awareness, policy support, access to finance and resources, landscape-level planning, and collaboration among stakeholders. By adopting and promoting agroforestry practices, it can foster resilient and productive farming systems that contribute to food security, poverty reduction, and sustainable development. Embracing agroforestry is not only a scientific solution but also an opportunity to create a more sustainable and inclusive future for agriculture.

Author Contributions: Conceptualization, N.S.D. and A.J.A.; methodology, S.S.U.; validation, A.J.A. and T.D.N.; formal analysis, N.S.D.; investigation, S.S.U.; writing—original draft preparation, N.S.D. and S.S.U.; writing—review and editing, A.J.A. and T.D.N.; supervision, A.J.A.; visualization, N.S.D. and S.S.U.; project administration, A.J.A. All authors contributed to the article and approved the submitted version.

Funding: This research received no external funding.

Acknowledgments: We would like to extend our gratitude to the technical team at the Agronomy Division of the Coconut Research Institute for their valuable contributions. We also want to express our thanks to the editor and two anonymous reviewers for providing constructive feedback and valuable comments.

Conflicts of Interest: The authors declare no conflict of interest.

References

Adams, Mark A., Judy Simon, and Sebastian Pfautsch. 2010. Woody legumes: A review from the South. *Tree Physiology* 30: 1072–82. [CrossRef]

Ajayi, Oluyede C. 2007. User acceptability of sustainable soil fertility technologies: Lessons from farmers' knowledge, attitude and practice in southern Africa. *Journal of Sustainable Agriculture* 30: 21–40. [CrossRef]

Albrecht, Alain, and Serigne T. Kandji. 2003. Carbon sequestration in tropical agroforestry systems. *Agriculture, Ecosystems & Environment* 99: 15–27. [CrossRef]

Atangana, Alain, Damase Khasa, Scott Chang, Ann Degrande, Alain Atangana, Damase Khasa, Scott Chang, and Ann Degrande. 2014a. Definitions and classification of agroforestry systems. In *Tropical Agroforestry*. Edited by Alain Atangana, Damase Khasa, Scott Chang and Ann Degrande. Dordrecht: Springer, pp. 35–47. [CrossRef]

Atangana, Alain, Damase Khasa, Scott Chang, Ann Degrande, Alain Atangana, Damase Khasa, Scott Chang, and Ann Degrande. 2014b. Phytoremediation in tropical agroforestry. In *Tropical Agroforestry*. Edited by Alain Atangana, Damase Khasa, Scott Chang and Ann Degrande. Dordrecht: Springer, pp. 343–51. [CrossRef]

Atapattu, Anjana J., Sumith H. S. Senarathne, Thilina Raveendra, Chaminda Egodawatta, and Sylvanus Mensah. 2017. Effect of short-term agroforestry systems on soil quality in marginal coconut lands in Sri Lanka. *Agricultural Research Journal* 54: 324–28. [CrossRef]

Ayyam, Velmurugan, Swarnam Palanivel, and Sivaperuman Chandrakasan. 2019. Agroforestry for livelihood and biodiversity conservation. In *Coastal Ecosystems of the Tropics—Adaptive Management*. Edited by Velmurugan Ayyam, Swarnam Palanivel and Sivaperuman Chandrakasan. Singapore: Springer Singapore, pp. 363–89. [CrossRef]

Banyal, Rakesh, Rajkumar, Manish Kumar, Rajender Kumar Yadav, and Jagdish Chander Dagar. 2018. Agroforestry for rehabilitation and sustenance of saline ecologies. In *Agroforestry: Anecdotal to Modern Science*. Edited by Jagdish Chander Dagar and Vindhya Prasad Tewari. Singapore: Springer Singapore, pp. 413–54. [CrossRef]

Beetz, Alice E. 2011. *Agroforestry: Overview*. Karnataka: ATTRA.

Bentrup, Gary, Jennifer Hopwood, Nancy Lee Adamson, and Mace Vaughan. 2019. Temperate agroforestry systems and insect pollinators: A review. *Forests* 10: 981. [CrossRef]

Beule, Lukas, Anna Vaupel, and Virna Estefania Moran-Rodas. 2022. Abundance, diversity, and function of soil microorganisms in temperate alley-cropping agroforestry systems: A review. *Microorganisms* 10: 616. [CrossRef]

Bhardwaj, Daulat Ram, Mansi R. Navale, and Sandeep Sharma. 2017. Agroforestry practices in temperate regions of the world. In *Agroforestry: Anecdotal to Modern Science*. Edited by Jagdish Chander Dagar and Vindhya Prasad Tewari. Singapore: Springer Singapore, pp. 413–54. [CrossRef]

Buck, L., S. Scherr, L. Trujillo, J. Mecham, and M. Fleming. 2020. Using integrated landscape management to scale agroforestry: Examples from Ecuador. *Sustainability Science* 15: 1401–15. [CrossRef]

Centeno-Alvarado, Diego, Ariadna Valentina Lopes, and Xavier Arnan. 2023. Fostering pollination through agroforestry: A global review. *Agriculture, Ecosystems & Environment* 351: 108478. [CrossRef]

Chavan, Sangram Bhanudas, Ram Newaj, Raza H. Rizvi, Ajit, Rajendra Prasad, Badre Alam, A. K. Handa, S. K. Dhyani, Amit Jain, and Dharmendra Tripathi. 2021. Reduction of global warming potential vis-à-vis greenhouse gases through traditional agroforestry systems in Rajasthan, India. *Environment, Development and Sustainability* 23: 4573–93. [CrossRef]

Chuma, Geant Basimine, Nadege Cizungu Cirezi, Jean Mubalama Mondo, Yannick Mugumaarhahama, Deckas Mushamalirwa Ganza, Karume Katcho, Gustave Nachigera Mushagalusa, and Serge Schmitz Serge. 2021. Suitability for agroforestry implementation around Itombwe Natural Reserve (RNI), eastern DR Congo: Application of the Analytical Hierarchy Process (AHP) approach in geographic information system tool. *Trees, Forests and People* 6: 100125. [CrossRef]

Dagar, Jagdish Chander, Sharda Rani Gupta, and Demel Teketay. 2020. *Agroforestry for Degraded Landscapes: Recent Advances and Emerging Challenges—Volume 1*. Singapore: Springer Singapore, p. 554. [CrossRef]

De Zoysa, Mangala. 2022. Scientific basis of agroforestry homegardens in Matara district, Sri Lanka: Present status and improvement needs. *Open Journal of Forestry* 12: 60–87. [CrossRef]

Den Herder, Michael, Gerardo Moreno, Rosa M. Mosquera-Losada, João H. N. Palma, Anna Sidiropoulou, Jose J. Santiago Freijanes, Josep Crous-Duran, Joana A. Paulo, Margarida Tomé, Anastasia Pantera, and et al. 2017. Current extent and stratification of agroforestry in the European Union. *Agriculture, Ecosystems & Environment* 241: 121–32. [CrossRef]

Dhyani, Shalini, Indu K. Murthy, Rakesh Kadaverugu, Rajarshi Dasgupta, Manoj Kumar, and Kritika Adesh Gadpayle. 2021. Agroforestry to achieve global climate adaptation and mitigation targets: Are South Asian countries sufficiently prepared? *Forests* 12: 303. [CrossRef]

Dissanayaka, Nuwandhya S., Lakmini Dissanayake, Shashi S. Udumann, Tharindu D. Nuwarapaksha, and Anjana J. Atapattu. 2023. Agroforestry—A key tool in the climate-smart agriculture context: A review on coconut cultivation in Sri Lanka. *Frontiers in Agronomy* 5: 1162750. [CrossRef]

Do, Hoa, Eike Luedeling, and Cory Whitney. 2020. Decision analysis of agroforestry options reveals adoption risks for resource-poor farmers. *Agronomy for Sustainable Development* 40: 1–12. [CrossRef]

Duffy, Colm, Gregory G. Toth, Robert P. O. Hagan, Peter C. McKeown, Syed Ajijur Rahman, Yekti Widyaningsih, Terry C. H. Sunderland, and Charles Spillane. 2021. Agroforestry contributions to smallholder farmer food security in Indonesia. *Agroforestry Systems* 95: 1109–24. [CrossRef]

Facheux, Charly, Zac Tchoundjeu, Divine Foundjem-Tita, Ann Degrande, and Mbosso Charlie. 2007. Optimizing the Production and Marketing of NTFPs. *African Crop Science Conference Proceedings* 8: 1249–54.

Fahad, Shah, Sangram Bhanudas Chavan, Akash Ravindra Chichaghare, Appanderanda Ramani Uthappa, Manish Kumar, Vijaysinha Kakade, Aliza Pradhan, Dinesh Jinger, Gauri Rawale, Dinesh Kumar Yadav, and et al. 2022. Agroforestry systems for soil health improvement and maintenance. *Sustainability* 14: 14221. [CrossRef]

FAO (Food and Agricultural Organization). 2022. *FAO and the Sustainable Development Goals—Achieving the 2030 Agenda through Empowerment of Local Communities*. Rome: FAO. [CrossRef]

Foster, Kristi, and Henry Neufeldt. 2014. Biocarbon projects in agroforestry: Lessons from the past for future development. *Current Opinion in Environmental Sustainability* 6: 148–54. [CrossRef]

François, Mathurin, Maria Carolina Gonçalves Pontes, Arthur Lima da Silva, and Eduardo Mariano-Neto. 2023. Impacts of cacao agroforestry systems on climate change, soil conservation, and water resources: A review. *Water Policy* 25: 564. [CrossRef]

Gikas, Georgios D., Vassilios A. Tsihrintzis, and Dimitrios Sykas. 2016. Effect of trees on the reduction of nutrient concentrations in the soils of cultivated areas. *Environmental Monitoring and Assessment* 188: 1–19. [CrossRef] [PubMed]

Griffiths, J., D. S. Phillips, Steve Compton, C. Wright, and L. D. Incoll. 1998. Responses of slug numbers and slug damage to crops in a silvoarable agroforestry landscape. *Journal of Applied Ecology* 35: 252–60. [CrossRef]

Haggar, Jeremy, Diego Pons, Laura Saenz, and Margarita Vides. 2019. Contribution of agroforestry systems to sustaining biodiversity in fragmented forest landscapes. *Agriculture, Ecosystems & Environment* 283: e106567. [CrossRef]

Hoffman, Brian M., Dmitriy Lukoyanov, Zhi-Yong Yang, Dennis R. Dean, and Lance C. Seefeldt. 2014. Mechanism of nitrogen fixation by nitrogenase: The next stage. *Chemical Reviews* 114: 4041–62. [CrossRef] [PubMed]

Hugenschmidt, Johannes, and Sonja Kay. 2023. Unmasking adaption of tree root structure in agroforestry systems in Switzerland using GPR. *Geoderma Regional* 34: e00659. [CrossRef]

Hussain, Zafar, Fahad Rasheed, Muhammad Ayyoub Tanvir, Zikria Zafar, Muhammad Rafay, Muhammad Mohsin, Pertti Pulkkinen, and Charles Ruffner. 2020. Increased antioxidative enzyme activity mediates the phytoaccumulation potential of Pb in four agroforestry tree species: A case study under municipal and industrial wastewater irrigation. *International Journal of Phytoremediation* 23: 704–14. [CrossRef] [PubMed]

Isaac, Marney E., Bonnie H. Erickson, S. James Quashie-Sam, and Vic R. Timmer. 2007. Transfer of knowledge on agroforestry management practices: The structure of farmer advice networks. *Ecology and Society* 12: 1–12. Available online: https://www.jstor.org/stable/26267879 (accessed on 20 July 2023). [CrossRef]

Jinger, Dinesh, Raj Kumar, Vijaysinha Kakade, Dhakshanamoorthy Dinesh, Gaurav Singh, Vinod Chandra Pande, P. R. Bhatnagar, B. K. Rao, Anand Kumar Vishwakarma, Dinesh Kumar, and et al. 2022. Agroforestry for controlling soil erosion and enhancing system productivity in ravine lands of Western India under climate change scenario. *Environmental Monitoring and Assessment* 194: 267. [CrossRef]

Jose, Shibu. 2009. Agroforestry for ecosystem services and environmental benefits: An overview. *Agroforestry Systems* 76: 1–10. [CrossRef]

Kareem, Idayah A., Michael F. Adekunle, Dorcas A. Adegbite, and Jubril Akanni Soaga. 2017. Evaluation of selected agroforestry practices and farmers' perception of climate change in Ogun State, Nigeria. *Forest Research Engineering: International Journal* 1: 9–16. [CrossRef]

Kassie, Geremew Worku. 2018. Agroforestry and farm income diversification: Synergy or trade-off? The case of Ethiopia. *Environmental Systems Research* 6: 1–14. [CrossRef]

Kaul, Meenakshi, Frits Mohren, and Vinay K. Dadhwal. 2010. Carbon storage and sequestration potential of selected tree species in India. *Mitigation and Adaptation Strategies for Global Change* 15: 489–510. [CrossRef]

Kiyani, Pilote, Jewel Andoh, Yohan Lee, and Don Koo Lee. 2017. Benefits and challenges of agroforestry adoption: A case of Musebeya sector, Nyamagabe District in southern province of Rwanda. *Forest Science and Technology* 13: 174–80. [CrossRef]

Klein, Alexandra-Maria, Ingolf Steffan-Dewenter, and Teja Tscharntke. 2003. Pollination of *Coffea canephora* in relation to local and regional agroforestry management. *Journal of Applied Ecology* 40: 837–45. [CrossRef]

Kommey, Randy, and Madeleine Fombad. 2023. Strategies for knowledge sharing among rice farmers: A Ghanaian perspective. *Electronic Journal of Knowledge Management* 21: 114–29. [CrossRef]

Kuyah, Shem, Sileshi G. Weldesemayat, Eike Luedeling, Festus K. Akinnifesi, Cory W. Whitney, Jules Bayala, Kuntashula Elias, Kangbeni Dimobe, and Paramu L. Mafongoya. 2020. Potential of agroforestry to enhance livelihood security in Africa. In *Agroforestry for Degraded Landscapes*. Edited by Jagdish Chander Dagar, Sharda Rani Gupta and Demel Teketay. Singapore: Springer Singapore, pp. 135–67. [CrossRef]

Lal, Rattan. 2016. Feeding 11 billion on 0.5 billion hectare of area under cereal crops. *Food and Energy Security* 5: 239–51. [CrossRef]

Lawson, Gerry, William J. Bealey, Christian Dupraz, and Ute M. Skiba. 2020. Agroforestry and opportunities for improved nitrogen management. In *Just Enough Nitrogen: Perspectives on How to Get There for Regions with Too Much and Too Little Nitrogen*. Edited by Mark A. Sutton, Kate E. Mason, Albert Bleeker, W. Kevin Hicks, Cargele Masso, Nandula Raghuram, Stefan Reis and Mateete Bekunda. Springer: Cham, pp. 393–417. [CrossRef]

Lee-Mader, E., M. Vaughan, and J. Goldenetz-Dollar. 2020. Agroforestry and cover cropping for pollinators. In *Towards Sustainable Crop Pollination Services: Measures at Field, Farm and Landscape Scales*. Edited by Barbara Gemmill-Herren, N. Azzu, Abram J. Bicksler and A. Guidotti. Rome: FAO, pp. 105–26.

Mahmud, Aliyu Ahmad, Abhishek Raj, and Manoj Kumar Jhariya. 2021. Agroforestry systems in the tropics: A critical review. *Agricultural and Biological Research* 37: 83–87.

Mathijs, Erik, and Erwin Wauters. 2020. Making farming systems truly resilient. *EuroChoices* 19: 72–76. [CrossRef]

Mbow, Cheikh, Pete Smith, David Skole, Lalisa Duguma, and Mercedes Bustamante. 2014. Achieving mitigation and adaptation to climate change through sustainable agroforestry practices in Africa. *Current Opinion in Environmental Sustainability* 6: 8–14. [CrossRef]

Meuwissen, Miranda P. M., Peter H. Feindt, Alisa Spiegel, Catrien J. A. M. Termeer, Erik Mathijs, Yann De Mey, Robert Finger, Alfons Balmann, Erwin Wauters, Julie Urquhart, and et al. 2019. A framework to assess the resilience of farming systems. *Agricultural Systems* 176: e102656. [CrossRef]

Mohammadi, Khosro, Gholamreza Heidari, Shiva Khalesro, and Yousef Sohrabi. 2011. Soil management, microorganisms and organic matter interactions: A review. *African Journal of Biotechnology* 10: 19840–49. [CrossRef]

Montagnini, Florencia, and Ruth Metzel. 2017. The contribution of agroforestry to sustainable development goal 2: End hunger, achieve food security and improved nutrition, and promote sustainable agriculture. In *Integrating Landscapes: Agroforestry for Biodiversity Conservation and Food Sovereignty—Advances in Agroforestry*. Cham: Springer, vol. 12, pp. 11–45. [CrossRef]

Muoni, Tarirai, Eric Koomson, Ingrid Öborn, Carsten Marohn, Christine A. Watson, Göran Bergkvist, Andrew Barnes, Georg Cadisch, and Alan Duncan. 2020. Reducing soil erosion in smallholder farming systems in east Africa through the introduction of different crop types. *Experimental Agriculture* 56: 183–95. [CrossRef]

Musa, Fazilah, Nor Asyirah Lile, and Diana Demiyah Mohd Hamdan. 2019. Agroforestry practices contribution towards socioeconomics: A case study of Tawau communities in Malaysia. *Agriculture and Forestry* 65: 65–72. [CrossRef]

Nair, Pradeep Kumar Ramachandran, B. Mohan Kumar, and Vimala D. Nair. 2021. *An Introduction to Agroforestry*. Cham: Springer International Publishing. [CrossRef]

Nath, Arun Jyoti, Rakesh Kumar, N. Bijayalaxmi Devi, Pebam Rocky, Krishna Giri, Uttam Kumar Sahoo, Raj Kumar Bajpai, Netrananda Sahu, and Rajiv Pandey. 2021. Agroforestry land suitability analysis in the Eastern Indian Himalayan region. *Environmental Challenges* 4: 100199. [CrossRef]

Ndlovu, Nicholas P., and Lars Borrass. 2021. Promises and potentials do not grow trees and crops: A review of institutional and policy research in agroforestry for the Southern African region. *Land Use Policy* 103: 105298. [CrossRef]

Nerlich, K., Simone Graeff-Hönninger, and W. Claupein. 2013. Agroforestry in Europe: A review of the disappearance of traditional systems and development of modern agroforestry practices, with emphasis on experiences in Germany. *Agroforestry Systems* 87: 475–92. [CrossRef]

Ngarava, Saul, Leocadia Zhou, Thulani Ningi, and Martin Munashe Chari. 2022. *Differentiated Intra-Household Food Utilisation in Raymond Mhlaba Local Municipality, South Africa*. World Sustainability Series; Edited by Walter Leal Filho, Marina Kovaleva and Elena Popkova. New York: Springer Science and Business Media Deutschland GmbH, pp. 87–106. [CrossRef]

Nuwarapaksha, Tharindu D., Nuwandhya S. Dissanayaka, Shashi S. Udumann, and Anjana J. Atapattu. 2023. Gliricidia as a beneficial crop in resource-limiting agroforestry systems in Sri Lanka. *Indian Journal of Agroforestry* 25: 12–18.

Octavia, Dona, Sri Suharti, Murniati, I. Wayan Susi Dharmawan, Hunggul Yudono Setio Hadi Nugroho, Bambang Supriyanto, Dede Rohadi, Gerson Ndawa Njurumana, Irma Yeny, Aditya Hani, and et al. 2022. Mainstreaming smart agroforestry for social forestry implementation to support sustainable development goals in Indonesia: A review. *Sustainability* 14: 9313. [CrossRef]

Pavlidis, George, and Tsihrintzis Vassilios. 2017. Pollution control by agroforestry systems: A short review. *European Water* 59: 297–301.

Pavlidis, George, Helen Karasali, and Vassilios A. Tsihrintzis. 2020. Pesticide and fertilizer pollution reduction in two alley cropping agroforestry cultivating systems. *Water, Air, & Soil Pollution* 231: 1–23. [CrossRef]

Platis, Dimitrios P., Christos D. Anagnostopoulos, Aggeliki D. Tsaboula, Georgios C. Menexes, Kiriaki L. Kalburtji, and Andreas P. Mamolos. 2019. Energy analysis, and carbon and water footprint for environmentally friendly farming practices in agroecosystems and agroforestry. *Sustainability* 11: 1664. [CrossRef]

Rahman, Syed Ajijur, Yusuf B. Samsudin, Kishor Prasad Bhatta, Anisha Aryal, Durrah Hayati, Muhardianto Cahya, Bambang Trihadmojo, Iqbal Husani, Sarah Andini, Sari Narulita, and et al. 2023. The role of agroforestry systems for enhancing biodiversity and provision of ecosystem services in agricultural landscapes in Southeast Asia. In *Agroforestry for Sustainable Intensification of Agriculture in Asia and Africa*. Edited by Jagdish Chander Dagar, Sharda Rani Gupta and Gudeta Weldesemayat Sileshi. Singapore: Springer Nature Singapore, pp. 303–19. [CrossRef]

Ramil Brick, Elisa S., John Holland, Dimitris E. Anagnostou, Keith Brown, and Marc P. Y. Desmulliez. 2022. A review of agroforestry, precision agriculture, and precision livestock farming—The case for a data-driven agroforestry strategy. *Frontiers in Sensors* 3: 998928. [CrossRef]

Riyadh, Z. A., Md. Ashadur Rahman, Satya Ranjan Saha, and M. I. Hossain. 2018. Soil properties under jackfruit based agroforestry systems in Madhupur tract of Narsingdi district. *Journal of the Sylhet Agricultural University* 5: 173–79.

Rosenstock, Todd S., Ian K. Dawson, Ermias Aynekulu, Susan Chomba, Ann Degrande, Kimberly Fornace, Ramni Jamnadass, Anthony Kimaro, Roeland Kindt, Christine Lamanna, and et al. 2019. A planetary health perspective on agroforestry in Sub-Saharan Africa. *One Earth* 1: 330–44. [CrossRef]

Sagastuy, Mauricio, and Torsten Krause. 2019. Agroforestry as a biodiversity conservation tool in the Atlantic forest? Motivations and limitations for small-scale farmers to implement agroforestry systems in north-eastern Brazil. *Sustainability* 11: 6932. [CrossRef]

Santoro, Antonio, Martina Venturi, Remo Bertani, and Mauro Agnoletti. 2020. A review of the role of forests and agroforestry systems in the FAO Globally Important Agricultural Heritage Systems (GIAHS) program. *Forests* 11: 860. [CrossRef]

Sarvade, Somanath, and Rahul Singh. 2014. Role of agroforestry in food security. *Popular Kheti* 2: 25–29.

Sarvade, Somanath, Rahul Singh, Heerendra Prasad, and Dasharath Prasad. 2014. Agroforestry practices for improving soil nutrient status. *Popular Kheti* 2: 60–64.

Shah, Shipra. 2023. Policy interventions for scaling up agroforestry in Fiji. *New Zealand Economic Papers* 57: 105–13. [CrossRef]

Sharma, Padma, Amarpreet Kaur, Daizy R. Batish, Shalinder Kaur, and Bhagirath S. Chauhan. 2022. Critical insights into the ecological and invasive attributes of *Leucaena leucocephala*, a tropical agroforestry species. *Frontiers in Agronomy* 4: 890992. [CrossRef]

Sharma, Rajni, Sanjeev K. Chauhan, and Abhishek M. Tripathi. 2016. Carbon sequestration potential in agroforestry system in India: An analysis for carbon project. *Agroforestry Systems* 90: 631–44. [CrossRef]

Simelton, Elisabeth S., Delia C. Catacutan, Thu C. Dao, Bac V. Dam, and Thinh D. Le. 2017. Factors constraining and enabling agroforestry adoption in Viet Nam: A multi-level policy analysis. *Agroforestry Systems* 91: 51–67. [CrossRef]

Smith, Matthew M., Gary Bentrup, Todd Kellerman, Katherine MacFarland, Richard Straight, and Lord Ameyaw. 2021. Windbreaks in the United States: A systematic review of producer-reported benefits, challenges, management activities and drivers of adoption. *Agricultural Systems* 187: 103032. [CrossRef]

Sollen-Norrlin, Maya, Bhim Bahadur Ghaley, and Naomi Laura Jane Rintoul. 2020. Agroforestry benefits and challenges for adoption in Europe and beyond. *Sustainability* 12: 7001. [CrossRef]

Staton, Tom, Tom D. Breeze, Richard J. Walters, Jo Smith, and Robbie D. Girling. 2022. Productivity, biodiversity trade-offs, and farm income in an agroforestry versus an arable system. *Ecological Economics* 191: e107214. [CrossRef]

Suroso, Johan Iskandar, Susanti Withaningsih, Deden Nurjaman, and Budiawati S. Iskandar. 2023. Bird population and bird hunting in the rural ecosystem of Cijambu, Sumedang, West Java, Indonesia. *Biodiversitas Journal of Biological Diversity* 24: 4470–84. [CrossRef]

Suzaki, Takuya, Emiko Yoro, and Masayoshi Kawaguchi. 2015. Leguminous plants: Inventors of root nodules to accommodate symbiotic Bacteria. *International Review of Cell and Molecular Biology* 316: 111–58. [CrossRef]

Thomas, George V., Vishnu Potti Krishnakumar, R. Dhanapal, and D. V. Srinivasa Reddy. 2018. Agro-management practices for sustainable coconut production. In *The Coconut Palm (Cocos nucifera L.)—Research and Development Perspectives*. Edited by K. U. K. Nampoothiri, Vishnu Potti Krishnakumar, P. K. Thampan and M. Achuthan Nair. Singapore: Springer, pp. 227–22. [CrossRef]

Tsegaye, Nigus Tekleselassie. 2023. The Impact of agroforestry practice on forest conservation and community livelihood improvement: A case of Buno Bedele zone of West Ethiopia'S Chora district. *Environmental and Sustainability Indicators* 17: 100229. [CrossRef]

Udawatta, Ranjith P., Lalith M. Rankoth, and Shibu Jose. 2019. Agroforestry and biodiversity. *Sustainability* 11: 2879. [CrossRef]

Xue, Zhang. 2023. Review and prospect of the promotion of agroforestry system technology in Southern collective forest areas. *International Journal of Engineering and Management Research* 13: 17–22. [CrossRef]

Improving the Diversity of Native Edible Plants and Traditional Food and Agriculture Practices for Sustainable Food Security in the Future

Permani C. Weerasekara and Angelika Ploeger

1. Introduction

By the year 2030, agriculture will have to provide the food and nutritional requirements of some eight billion people (Godfray et al. 2010). These include eradicating hunger, improving access to food, ending all forms of malnutrition, promoting sustainable agriculture, and preserving food diversity (Godfray et al. 2010). Thus, food security or, even better, food sovereignty, is one of the challenges in this world (Boyer 2010). Simultaneously, research and development are focused on improving the productivity of a small number of existing crops that will improve global food production instead of increasing the diversity of crops. The result is the loss of agrobiodiversity (Shiferaw et al. 2011). This results in a loss of agricultural biodiversity, which, in turn, results in a food industry that is more vulnerable to abiotic and biotic stresses and at an increased risk of catastrophic losses (Shelef et al. 2017). Increasing global food production is needed to address these challenges, which cannot be achieved by expanding industrial agriculture by converting land into an environmentally degraded environment and biodiversity (Chappell and LaValle 2011). The use of wild plants and the development of new crops will, therefore, help us diversify global food production and better adapt to the diversity and changing living conditions of populations (Shelef et al. 2017).

Additionally, the manifestations of global, climate, environmental, behavioral, and technological changes underscore the need to improve food production in ways that minimize the negative impacts of sustainability on the ecosystems we believe to be sustainable (Godfray and Garnett 2014). Indigenous ecological agriculture can restore the health of land and people through five processes that activate and connect all life. The five processes are the flow of energy absorbed by plants through photosynthesis, the soil-mineral cycles that provide nutrients for life, the water cycles, the biosphere, our ecological relationships that create animal communities, and the relationships between people and land, including the genes (Massy 2020). As part of these vital connections, we can also benefit from relearning how to use native plants as sources of healthy food and other products, with a focus on environmental issues. Shelef et al. (2017) argue that the use of indigenous plants in local food production will help create more sustainable agriculture. Native food production has recently received more attention, while the use of native plants in local food production has received less attention. According to the UN Food and Agriculture Organization (FAO), more than 90% of the calories consumed by humans come from just 30 plant species (Hammer et al. 2003). Humans only cultivate about 150 of the world's estimated 30,000 edible plant species, and only 30 plant species make up most of our food (Shelef et al. 2017; Sethi and Plummer 2015). For some of these species, genetic diversity decreases as the number of varieties sold decreases (Shelef et al. 2017).

In any particular culture, no matter how popular they are, traditional foods have long been consumed and are considered to be an expression of history and lifestyle (De Soucey 2010). However, the intergenerational dissemination of knowledge about indigenous and traditional foods is now limited due to changing lifestyles, there being fewer knowledge holders, and there being fewer flora and fauna resources. Therefore, new crops and weeds of an indigenous origin can diversify global food production through commercial practices, as well as the use of traditional knowledge and allow better local adaptation to human habitation in different environments.

Regarding this matter, this study focused on Sri Lanka. In this article, we consider Sri Lankan edible plants, their values and advantages, and the barriers to using local traditional food plants, knowledge, and their uses. Sri Lankan traditional diets are richly diverse (Weerasekara et al. 2018) and offer various health and nutritional benefits, including protection from noncommunicable diseases (NCDs) and micro nutrition deficiencies. As well, Sri Lanka is an island rich in biodiversity and has approximately identified 3368 plant species belonging to 1294 genera and 132 families (Rajapaksha 1998). About 800 plants are endemic to Sri Lanka (Rajapaksha 1998). Many of these plants have been used in the past to build a healthy rural community under the precise guidance of local traditional healers, elders, and indigenous communities. Local and traditional foods in Sri Lanka have a long history, and unique traditions have existed for thousands of years (Perera 2008). The traditional diet in Sri Lanka is closely linked to nutritional, health, and therapeutic arguments regarding food ingredients and preparation methods (Weerasekara et al. 2018). These plant species have unique therapeutic and nutritional properties that can solve acute local health problems (Weerasekara et al. 2020). This article examines the local plants and traditional nutritional knowledge. This article helps to improve the diversity of native edible plants and traditional food and agriculture practices for sustainable food security in the future.

2. Materials and Methods

2.1. Study Area

The study is part of a collaborative project investigating food system commercialization and hidden hunger and malnutrition in Sri Lanka (Weerasekara et al. 2018; Weerasekara et al. 2020). Based on the feasibility and previous research experience carried out in Sri Lanka in 2016–2019, the agro-environmental zone for rural areas was selected for this project. The district of Anuradhapura in the north–central province of Sri Lanka was chosen as the study area. The people in this area are mainly engaged in agriculture to earn their main source of income. According to our recent study, the people in this area had rich dietary diversity in the past (Weerasekara et al. 2018).

2.2. Study Sample

This research used the combined concept of ethnographic and sociological study approaches. Information and data were collected through field interviews and historical references. Oral histories are a collection of stories and the reminiscences of a person or persons who have first-hand knowledge of any number of experiences (Ritchie 2014; Thompson 2017). Therefore, the interviewers carried out discussions with older people over the ages of 70–90 years (n = 50) who had experience with traditional knowledge and with the subject experts (n = 10). The expert interviews were different to each other in terms of the questions and subjects. The experts were selected based on the nature of the research questions. Furthermore, there were open interviews and questions with small-scale farming households (n = 20). These were mainly incorporated to capture any historical production-related changes in the traditional food system and dietary patterns in Sri Lanka. Food security Participants were asked questions about their access to traditional foods. A multi-step coded question asked participants about their perceptions of their current traditional food consumption in comparison to 50 years ago.

2.3. Data Analysis

The interviews were conducted in Sinhala, and then translated into English. The interview data were transferred to an Excel datasheet. MAXQDA 2018 was used for coding the interviews. The interview data were also transferred to the datasheet before conducting statistical analysis. The interview questions were slightly altered based on research areas and position. However, it remained connected to the research objectives. The Sri Lankan food consumption table was used to calculate nutritional analysis.

3. Results

Historical references showed that, in ancient Sri Lanka, there were no farmers as they are defined today because people never owned farms or farmed for money. Agriculture was not a revenue-generating process, and at the same time, it was not considered to be a business or an industry. Agriculture was essentially everybody's service, and it was the public's responsibility to use and maintain the land for the sake of the nation (Weerasekara et al. 2018). Therefore, Sri Lanka is an island with high biodiversity, and access to food was not a problem. This is well documented in the old chronicles by Robert Knox (1983) and Emerson Tennent (1860). Most of the foods that the Sri Lankan elders enjoyed were not grown by them. They were found everywhere, growing naturally (Knox 1983). This was confirmed by the elderly people in this area.

Eighty-five species were identified, dominated by vegetables, fruits, legumes, roots/tubers, herbs, wild mushrooms, spices, and cereals, in this study. The findings show that Sri Lankan people had diverse crops in their home gardens. These findings suggest that traditional food with high agrobiodiversity contributed more toward food security among Sri Lankans in the selected sites. These findings suggest that traditional food with high agrobiodiversity contributed more toward food security among Sri Lankans in the selected sites. The finding of the study showed many traditional Sri Lankans have always been concerned about the type of food that they choose, including the quantity and quality of their food. Also, food security and food availability in traditional Sri Lanka were higher. Food was consumed according to the type of person (child, adult, elder), physiology (sick, pregnant, nursing), degree of activity (less active, energetic), and the type of meal (breakfast, lunch, dinner).

Based on the data, the results can be subdivided into traditional food plants, traditional food preparation methods, traditional food preservation, food security, and traditional agricultural practices.

3.1. Traditional Native Food Plants

According to the results, traditional native food plants can be categorized as green leafy vegetables, vegetables, fruits, grains, Pulses, roots, and tubers. Traditional food plants can be divided in two. One group represents the use of native plant species, which often have not been studied or commercialized. Another one is the commercial use of food plant species. These, not well-known plants have become an essential part of Sri Lanka's rural lifestyle and enhanced the country by providing people's daily therapeutic supplements for a healthy lifestyle at no extra cost.

3.1.1. Green Leafy Native Vegetables

Sri Lankan people utilized a large number of plant species to meet their food needs. Green leafy vegetables of various native plants with good health properties enrich their diet with fiber and micronutrients. The Sri Lankan diet consisted of green leaves, which they used in many ways. In particular, green leaves were the main source of vitamins and other therapeutic values. Green leafy vegetables can be classified into: (1) Cultivated vegetable leaves (many cultivated plant leaves are consumed) such as cassava leaves, pumpkin leaves, and beetroot leaves. (2) Semi-wild vegetables that grew as wild shrubs. But now, they are protected when they grow in home gardens. (3) Wild leafy vegetables that grew in forest areas. Most of the green vegetables and leaves listed contain high amounts of minerals and vitamins that play a significant role in boosting people's immune systems. These foods are still often available in forest areas or home gardens. Unfortunately, these plants are often not consumed, studied, or commercialized. Nineteen green leafy vegetable species were identified in this study (see Table 1). The historical references and oral interviews showed that some of the traditional Sri Lankan green leafy foods have many nutritional benefits.

Some of the leaves are less popular, but have a high nutrient value, and people can use and cultivate them. These are the examples for that: Bata Krilla (*Erythroxylum moonii*), Kara Kola (*Canthium coromandelicum*), Pothupala (*Aniseia martinicensis*), Genda (*Portulaca oleracea*), Pitawakka (*Phylanthus debilis*), and Kalu habarala (*Maba buxifolia*).

Table 1. Leafy green vegetables in the study area.

	Food Name	Pictures	Food Used	Nutritional Value/Health Value	Current Situation
1.	Pennywort (Wal-Gotukola)		The whole plant is used	This is an herb that grants longevity and is used in traditional medicine to prevent many diseases. This will help control memory deficit and self-heating problems. It also increases collagen production, which might be helpful for wound healing. It contains many vitamins and minerals (B, K, calcium, zinc, and magnesium).	Often consumed and commercialized
2.	*Alternanthera sessilis* (Weda mukunuwenna)		Young shoots and leaves	Mukunuwenna is traditionally used as herbal medicine to treat internal ailments. Mukunuwenna is used for simple stomach disorders, diarrhea, dysentery, and as a plaster for diseased or wounded skin. It contains high amount, of lipids, amino acids, vitamin A, protein, and calcium.	Rarely consumed
3.	Green milkweed climber (Aguna Kola)		There are two types of aguna leaves: kiri aguna, which is not bitter, and thiththa aguna, which is bitter	This leaf is used in traditional medicine and prevents many diseases. These leaves are detoxifiers, Lacto purifiers, and enhancers.	Rarely consumed; can buy it at some markets

Table 1. *Cont.*

	Food Name	Pictures	Food Used	Nutritional Value/Health Value	Current Situation
4.	Canereed (Thebu Kola)		Young leaves	This leaf is used in traditional medicine and prevents many diseases. The leaf of the plant helps in producing insulin. It controls blood sugar levels. This plant is full of antioxidants, which is good for fatty livers and kidneys. It is a rich source of nutrients, high in dietary fiber, and rich in folate, ascorbic acid, Mg, and, vitamins K.	Rarely consumed; can buy it at some markets
5.	Quail grass, *Celosia argentea* (Kirihanda)		Leaves and young steams	This leaf is used in traditional medicine and prevents many diseases. It is mainly used to treat swollen eyes, high blood pressure, nosebleeds, and other conditions associated with liver fire rising.	Rarely consumed
6.	Littlebell (Thal Kola)		Leaves	This leaf is used as a traditional medicine and prevents many diseases, such as kidney diseases, gastritis, and asthma.	Rarely consumed
7.	Snot berry (Lolu Kola)		Leaves and fruits	This leaf is used in traditional medicine and prevents many diseases such as urinary infections and bronchial infections.	Rarely consumed

Table 1. *Cont.*

Food Name	Pictures	Food Used	Nutritional Value/Health Value	Current Situation
8. Slender amaranth (Kuura Thampala)		Leaves and young stems	This leaf is high in iron and vitamins B and C. Used for gastritis, eaten cold, and good for the eye. Roots have been taken as medicine for worm diseases.	Rarely consumed; can buy it at some markets
9. Sickle Senna (Thora leaves)		Leaves	It is used to treat warm diseases, asthma, constipation, hemorrhoids, ringworms dysentery, eye diseases, liver disorders, and skin diseases. This will help to purify the blood and is good for the heart.	Rarely consumed
10. *Cardiospermum halicacabum* (Welpenela)		Leaves and young stems	Welpenela is used to treat dysentery, rheumatoid, arthritis, back pain, and hernia conditions and is an Ayurvedic medicine.	Rarely consumed; can buy it at some markets

Table 1. *Cont.*

	Food Name	Pictures	Food Used	Nutritional Value/Health Value	Current Situation
11.	Ivy Gourd (Kowakka)		Leaves	This plant is used for mellitus, liver disorders, gonorrhea, menorrhagia, wounds, leucorrhea, swelling, cough, polyurea, asthma, fevers, skin diseases, diabetes, jaundice, and anemia.	Rarely consumed; can buy it at some markets
12.	Horse Purslane (Sarana)		Leaves	This plant is a blood purifier; it treats many diseases from skin allergies to constipation. It is a good source of vitamin C, potassium, magnesium, and iron. It contains rich amounts of nutrients, antioxidants, vitamins, minerals, fiber, and fatty acids, and is a rich source of omega-3 fatty acids.	Rarely consumed; can buy it at some markets
13.	*Aporosa lindleyana* (Kebella)		Leaves	This plant has antioxidant, anti-amylase, and lipid-lowering properties. Used for urine diseases and eye diseases.	Rarely consumed

Table 1. *Cont.*

	Food Name	Pictures	Food Used	Nutritional Value/Health Value	Current Situation
14.	Water spinach (Kankun)		Leaves	Water spinach has a lot of nutrients to offer. It can reduce blood pressure, give immunity from cancer, improve vision, boost immunity, and treat skin diseases. It is a rich source of various vitamins, minerals, and antioxidants. It contains abundant quantities of water, iron, vitamin C, potassium, vitamin A, and other nutrients.	Commonly consumed
15.	Climbing day flower (Girapala)		Leaves	This plant is used to treat eye diseases, boils, burns, groin swellings, and pruritus.	Rarely consumed
16.	Ceylon-spinach (Gas Nivithi)		Leaves	This plant is used to treat lung diseases, male infertility, spleen disorders, impotence, diarrhea, loss of energy, and dysmenorrhea. It is rich in antioxidants, particularly beta carotene and lutein, which are naturally occurring chemicals that help keep your cells from aging. It contains vitamin B9, iron, calcium, copper and magnesium.	Rarely consumed

Table 1. *Cont.*

	Food Name	Pictures	Food Used	Nutritional Value/Health Value	Current Situation
17.	*Clitoria ternatea* (Katarolu)		Leaves	This plant is used to treat liver disorders, anasarca, swollen joints, ascites, the enlargement of abdominal viscera, dyspepsia, and the irritation of the bladder.	Rarely consumed
18.	Sweet broom (Wal koththamalli)		Leaves	This plant is used to treat diabetes mellitus, liver diseases, menstrual, parasitic infections, ear disease, eye diseases, venereal disease, stomach ailments, vomiting, leprosy, and Noso pharyngeal infection.	Rarely consumed
19.	Cluster tree (Attikka)		Leaves	This plant is used to treat diabetes mellitus, swellings, wounds, gonorrhea, burning sensation, rectal prolapse, pains, oral diseases, dysentery, skin diseases, menorrhagia, excessive thirst, and leucorrhoea.	Rarely consumed

Source: Authors' compilation based on data from Rajapaksha (1998); Jayaweera (1980); De Fonseka and Vinasithamby (1971); BFN Project Sri Lanka (2009); Biodiversity for Food and Nutrition (2018).

3.1.2. Vegetables and fruits

Ninteen species of native fruits and vegetables were identified in this study (see Table 2). Traditional Sri Lankans believed that eating vegetables daily is important for good health. Vegetables provide vitamins, minerals, and other essential nutrients, such as antioxidants and fiber. Vegetables provide the nutrients necessary for the well-being and maintenance of the human body. This reflects the growing need to promote less-exploited crops, especially conventional crops, to meet the food needs of Sri Lanka's growing population. The villagers defined "naturally grown vegetables" as edible plants grown without special care, without the use of pesticides, herbicides, and fertilizers, and that are not grown using commercial seeds or for commercial purposes (Perera 2008). The vegetables were picked daily around the houses, preferably before cooking. They also mentioned that freshly picked vegetables made their food tastier.

Table 2. Sri Lankan traditional vegetables and fruits.

Wild Fruit	Local Vegetables
Mora (*Dimocarpus longan*) This fruit is a good source of B vitamins and vitamin C. This fruit is used for insomnia, cardiovascular diseases, and diabetes mellitus. It contains fair amounts of minerals like phosphorus and copper.	**Thumba Karawila/Spine gourd (*Momordica dioica*)** This vegetable is native to natural forest areas, especially in high altitudes. This vegetable is eaten locally as a vegetable that has high nutritional and medicinal value. This vegetable cures stomach problems, diarrhea and asthma, leprosy, bronchitis, heart diseases, and urinary problems.
Weera (*Draperies septaria*) The fruits that are slightly astringent in taste are borne at the leaf axils and scattered profusely on all branches. This fruit is freely available in the forest. It is chemical-free and contains the nutritious value.	**Hal (*Vateria copallifera*)** This vegetable is endemic to Sri Lanka. The flour obtained from fruits is used in the preparation of traditional food. This food was used for diarrhea and diabetes.
Madan (*Syzygium cumini*) Barks and seeds are used to treat diabetes and the leaf juice is used for gingivitis conditions. The fruit contains iron and ascorbic acid. This fruit is used for heart and liver problems.	**Nelum Ala/Lotus root (*Nelumbo nucifera*)** It is rich in nutrients and has a lot of beneficial effects. Health benefits of eating lotus root include reduction in cholesterol, improvement in digestion, lower blood pressure, and also helps to boost the immunity system.

Table 2. *Cont.*

Wild Fruit	Local Vegetables
Palu (*Manikara hexandra*) This fruit is used to relieve burning sensation and anorexia while the bark is used for odontopathy.	**Wal del (*Artocarpus nobili*)** This plant is endemic to Sri Lanka. Waldel seeds are good for asthma patients. Wal del oil from seed was used in traditional medicine.
Eraminiya (*Ziziphus napeca*) This fruit is used for fever, dysentery, and loss of appetite. People also use this fruit for constipation, diabetes, aging skin, high cholesterol, insomnia, and many other conditions, but there is no good scientific evidence to support these uses. This fruit is freely available in the forest and it contains a high nutritional value.	**Kekatiya (*Aponogeton crispus*)** This plant is native to Sri Lanka. This plant is a good source of nutrition. These flowers are consumed as vegetables. This plant is a good source for burning sensations in the body, heart disease, and diabetes. Traditional men gave to pregnant women.
Karamba (*Carissa* spp.) It prevents excessive secretion of bile by the liver and is used for Ayurvedic medicine. This fruit is also rich in vitamin C, vitamin A, calcium, phosphorus, and iron.	**Emberella (*Spondias dulcis*)** This fruit is used for high blood pressure while leaves are used to cure mouth sores.
Veralu/Ceylon olive (*Elaeocarpus serratus*) This fruit is rich in minerals, vitamins, fiber, and valuable antioxidants. It possesses anti-inflammatory, antibiotic, antianxiety, analgesic, antidepressant, and antihypertensive properties. It is used in rheumatism and is an antidote for poison.	**Ash Pumpkin/Puhul (*Benincasa hispida*)** The fruit contains a fixed oil, starch, resin, proteins, vitamins B and C. It is used for insanity, epilepsy and other nervous diseases. The cortical part of the fruit is given to diabetics.
Katuanoda (*Annona muricata*) It has strong anti-cancer effects. Katuanoda is high in carbohydrates, especially fructose and It has large amounts of vitamins C, B1, and B2. It has an antioxidant known to boost immune health.	**Drumstick/Murunga (*Moringa oleifera*)** The fruits contain energy, iron, moisture, protein, fats, carbohydrates, calcium, phosphorus, carotene, thiamine, riboflavin, niacin, vitamins C and B. It was used for insanity, epilepsy and other nervous diseases. The cortical parts of the fruit are given for diabetes. Leaves and antidote bark of the tree are used in food preparation.

Table 2. *Cont.*

Wild Fruit	Local Vegetables
Damuna (*Grewia tiliifolia*) This fruit is freely available in the forest. It is chemical-free and contains the nutritious value.	**Kohila (*Lasia spinosa*)** Kohila is used as a vegetable. There are several varieties of kohila such as kiri kohila, well kohila, guru kohila, kalu kohila and goda kohila. The tubers, roots and leaves are used as medicine. It contains lots of fibre, calcium and vitamin C.
Beli (*Aegle marmelos*) It is a rich source of a variety of nutrients that are useful for human health since it includes several vitamins and minerals. It contains various nutrients like beta-carotene, protein, riboflavin, vitamin C, vitamin B1, and B2. It has anti-fungal, anti-parasitic, and anti-viral properties that help prevent bacterial or viral infections and skin issues. It helps flush out all the harmful toxins from the body, purify the blood, and thus prevent kidney and liver issues.	

Source: Authors' compilation based on data from BFN Project Sri Lanka (2009); Biodiversity for Food and Nutrition (2018).

The traditional diet consists of a large range of vegetables and fruit extracts of different species prepared in different ways. Jackfruit and breadfruit, which are found all over the island, provide many edible components. In Sri Lanka, jackfruit is consumed as a vegetable as well as a fruit. Young jackfruit seeds are high in fiber and have many health benefits, such as high levels of protein and vitamins (Swami et al. 2012), and they boost people's appetite when they are boiled, fried, or cooked. Also, different types of bitter gourd (fence, snake, crested, bitter, and bottle), gourd (gray gourd, Malay, button, Arjuna, Samson, Mimini, Ruhuna, and Janani), melon, bean (long, French, winged, and wide), eggplant, luffa, gray bananas, tomatoes, women's fingers, beets, drumsticks, radishes, leeks, banana flowers, and cucumbers are some of the most popular traditional vegetables. Sri Lankans also consume a variety of young fruits, such as bananas, mangoes, pineapples, and papayas, as vegetables. Additionally, different variations of ripe fruits, such as mango, pineapple papaya, passion fruit, sugar apple, durian, rambutan, mangosteen, wood apple, bael fruit, avocado, different kinds of banana ("kolikuttu", "seeni kesel", "ambul kesel"," suwadal", "puuwalu", "rathkehel", "ambun", etc.) and jackfruit, were common desserts that accompanied main meals.

In addition, the villagers defined "naturally grown vegetables" as edible plants grown without special care, without the use of pesticides, herbicides, and fertilizers, and those that are not grown using commercial seeds or for commercial purposes (Perera 2008). These vegetables were picked daily around houses, preferably before cooking. They also mentioned that freshly picked vegetables made their food tastier. Some local fruits, such as "beli", "masan", "mora", "himbutu", "nelli", "katu anoda", "veralu"(ceylon olive and "lavulu", are rapidly disappearing. More than 100 species of wild fruits have been recorded in Sri Lanka. Many of the selected wild fruits and vegetables are listed below (Table 2).

3.1.3. Grains and Cereal

Fourteen spices of traditional grains were found in this study (Table 3). Rice has been a staple food and a major carbohydrate source in the Sri Lankan diet since ancient times. Rice was consumed three times a day. Different varieties of rice were served to different people, such as pregnant and lactating women, sick people, monks, and children (Perera 2008; Weerasekara 2013). Pregnant mothers, for example, were given "Ma Vee" varieties and other varieties were fed "Heenati" to infants and adults who could not easily digest them. There are reports that there were more than 2442 different types of rice in ancient Sri Lanka (Perera 2008; Weerasekara et al. 2018). These traditional varieties have good nutritional benefits. In addition to rice, there were many types of grains in Sri Lanka.

3.1.4. Roots and Tubers

The study found five types of roots in this study area (see Table 4). Starchy roots and tubers are plants that originate from diversified botanical sources that store edible starch in underground stems, roots, corms, bulbs and tubers. After grains, starchy root and potato crops are important as a global carbohydrate source. In addition, various tubers and roots provided the carbohydrates in the Sri Lankan diet. The edible Dioscorea and colocasia species were traditionally the most popular in Sri Lanka. There are more than 93 varieties of roots and tubers crops in Sri Lanka, but despite their abundance, their use has been declining in recent decades. Starchy root and tuber crops are important components of the human diet. There are several roots and tubers of different species that produce great biodiversity even within the same geographical location. Starchy roots and tubers have been a part of dietary choices since ancient times.

Table 3. Most common traditional grains in Sri Lanka.

Type of Grains	Original Name	Nutritional Benefits
Rice (*Oryza sativa* L.)	Suwadel (*Oryza sativa*)	Flavorful white rice with a milky taste. This rice has more nutritional benefits than other rice varieties. This rice is fortified with vitamins, minerals, antioxidants and a very low glycemic index (GI) compared to other varieties of white rice.
	Kalu Heenati (*Oryza sativa*)	This rice is rich in minerals, zinc and iron. This is a nutritious meal for breastfeeding mothers. The high fiber content of this rice controls bowel movements and constipation. Porridge made from this rice variety is used as a medicine for snake bites. It is highly recommended for patients with high cholesterol, diabetes and diarrhea. Improves sexual energy and physical strength.
	Kuruluthuda (*Oryza sativa*)	This rice is a highly nutritious old type of rice. This rice has a rich flavor rich in protein and fiber. This variety is used to improve bladder function and is used in medicine and cosmetic treatments to reduce diabetes and cardiovascular disease. This is a high antioxidant.
	Madathawalu	This rice is high in protein, minerals and fats compared to much other rice and is rich in iron, zinc, potassium and antioxidants. Due to its many medicinal and nutritional properties, rice is important in the treatment of diabetes and oxidative stress.
	Pachchaperumal	It is rich in nutrients and protein and is considered an excellent option for the daily diet. It would also be part of a good diet for people with diabetes and cardiovascular disease. In traditional Sinhalese culture, pachchaperumal has long been considered divine rice.

Table 3. *Cont.*

Type of Grains	Original Name	Nutritional Benefits
Finger Millet/Kurakkan (*Eleusine coracana*)	Bala Kurakkan	Kurakkan is an amazing grain variety that has been grown in Sri Lanka for thousands of years. It is high in calcium, iron, magnesium and fiber content. It has beneficial properties for the heart and severity of asthma and reduces the frequency of migraine attacks. It may help lower blood pressure and reduce the risk of heart attack, especially in older people with atherosclerosis or diabetic heart disease.
	Wadimal Kurakkan	
Foxtail millet /Thanahaal (*Setaria italica*)	Ran, Kaha and Kalu Thanahaal	It is highly nutritious benefits. It is used as a medicine for snake bites. This grain is used for cooling the body.
Water Lily seeds (*Nymphaeaceae*)	Olu Haal	This grain has high vitamins and energy. It is used as a medicine for blood pressure and heart attack, fatty liver, urine problems, kidney diseases and piles (hemorrhoids).
Kodu Millet (*Paspalum scrobiculatum*)	Amu	Whole grains are high in fiber and high in protein. Regular consumption of this grain is beneficial for postmenopausal women who suffer from metabolic diseases such as cardiovascular disease, high blood pressure and high cholesterol and are at increased risk of colon cancer.
Proso Millet (*Panicum miliaceum*)	Meneri	Millet is rich in soluble and insoluble dietary fiber. The insoluble fiber in millet is called "probiotic", which means it supports the good bacteria in your digestive system. This type of fiber adds a large amount to the stool, maintaining it regularly and also reducing the risk of colon cancer.
Sorghum (*Sorghum bicolor*)	Idal Iguru	This is a nutritious grain. It is rich in vitamins and minerals. It is an excellent source of fiber, antioxidants and proteins.

Source: Authors' compilation based on data from Rajapaksha (1998); Jayaweera (1980); De Fonseka and Vinasithamby (1971); BFN Project Sri Lanka (2009); Biodiversity for Food and Nutrition (2018).

Table 4. Type of commonly used roots and tubers in traditional Sri Lanka.

Type of Roots and Tubers	Description
Katu Ala (*Dioscorea esculenta*)	This is a rare food that grows spontaneously in the forests and bushes of Sri Lanka. Traditional people burned this food in the fire and ate them. These yams were a staple of the traditional diet in those days. The effort of digging is eaten by the villagers themselves, and coming to market is rare.
Elephant yam/Raja Ala (*Amorphophallus campanulatus*)	Raja ala is known for its nutrient content, for example, protein, minerals and vitamin C. It is perfect for strengthening the immune system and is also known as an anti-inflammatory and serves as an excellent antibody against stomach and intestinal cramps.
Udala (*Dioscorea bulbifera*)	It has high nutritional content and high antioxidant capacity. This tuber was consumed in cases of hemorrhoids, hormonal deficiency, hysteria, sterility, nervous excitability, syphilis, cancer and abdominal pain.
Angili Ala (*Dioscorea alata*)	It has high micronutrients and minerals and high antioxidant capacity.
Kukulala (*Dioscorea esculenta*)	It has high nutritional content and high antioxidant capacity.

Source: Authors' compilation based on data from Rajapaksha (1998); Jayaweera (1980); De Fonseka and Vinasithamby (1971); BFN Project Sri Lanka (2009); Biodiversity for Food and Nutrition (2018).

3.1.5. Pulses and Legumes

Legumes play an important role as a source of protein in the Sri Lankan diet (Sherasia et al. 2018; Ofuya and Akhidue 2005). The health and nutritional benefits of pulses are remarkable. Therefore, they are considered to be a good source of nutrients to combat malnutrition in developing countries such as Sri Lanka. There are many species and varieties of legumes with different carbohydrates, proteins, minerals, vitamins, fiber, and other bioactive compounds. Sri Lanka is a valuable repository of agrobiodiversity and is rich in genetic diversity. The pulses grown in Sri Lanka reflect this value, and there are almost 100 varieties in Sri Lanka (Helvetas Sri Lanka 2001). The most common examples are cowpeas, peas, horse gram, black gram, lentils, common beans, and winged beans. The nutritional value, health benefits, and therapeutic effects of these varieties vary from species to species. Furthermore, genetically modified legume varieties are more tolerant of harsh environmental conditions. Therefore, it is essential to identify cereals with a high nutritional value and high level of tolerance to adverse environmental conditions, pests, and diseases.

3.1.6. Herbs and Spices

Sri Lanka was known as the "Pearl of the East" and the "Spice Island" in ancient times. The Portuguese, the Dutch, and finally, the British were drawn to Sri Lanka because of their passion for spices. Sri Lanka has a variety of plant-based spices (flavors) that have incredible health benefits, most of which we can use. As for the herbs used in everyday life, they can be used as medicines for many minor ailments. For example, if someone has a stomachache, they can roast cumin and drink it like coffee. In addition, goraka (*Garcinia cambogia*), consumed in moderation, can reduce cholesterol levels. It prolongs the life of various herbs and spices and adds flavor. All herbs and spices used in traditional Sri Lankan cuisine have antifungal, antimicrobial, bactericidal, antifungal, and/or antifungal properties, and anti-inflammatory and anti-diabetic properties have been reported (Weerakkody et al. 2010). Commonly used spices are listed in the following table (see Table 5).

Table 5. Commonly used spices in Sri Lanka.

Food Name	Used of Food and Benefits
Cinnamon/Kurudu (*Cinnamomum verum*)	It contains antioxidant, anti-inflammatory, anti-diabetic, antimicrobial, anticancer, hypolipidemic, cardiovascular, anti-inflammatory properties. Traditionally, people used it to treat toothache and improve the food's taste.
Cardemon/Enasal (*Elettaria cardamomum*)	It is commonly used as a breath freshener. Cardamom has been shown to have antioxidant, anti-inflammatory, antihypertensive, cholesterol-lowering, and blood sugar-lowering effects.
Cloves/Karabunati (*Syzygium aromaticum*)	It is commonly used to treat stomach aches and toothache and as a swelling worm treatment.
Coriander	It is used to treat colds, fevers, urine infection, and pains.
Curry leaf	It is used to counteract snake venom, and it treats stomach aches, high blood pressure, and a loss of appetite and controls cholesterol levels.
Gamboge/Goraka (*Garcinia morella*)	It is used to treat stomach ailments, foot worms, and fungal problems and swelling and weight loss.
Lemon Grass	It is used to treat allergies; the stalk is worn as an earring.

Table 5. *Cont.*

Food Name	Used of Food and Benefits
Fennel	It is used to treat stomach aches and grips caused by water.
Cumin	It is used to treat stomach aches and diarrhea.
Mustard	It is used to treat swelling and for traditional rituals.
Nutmeg/Mace	It is used to treat stomach aches and vomiting; it is included in wine and cakes; it is eaten with betel leaves.
Turmeric	It is used as a cosmetic, for the treatment of lizard bites, broken bones, pimples, sores, measles, chickenpox, and broken bones, and tincture.
Pepper	It is used to treat stomach aches and indigestion.
Fenugreek	It is good for pregnant women. It is used to treat stomach aches and rheumatic pains; it is applied as a treatment for the hair.
Garlic	It is used to treat rheumatic discomforts, stomach aches, heart problems, and most other ailments for pregnant mothers.
Ginger	It is used to treat colds, fevers, stomach aches, swellings, coughs, and sore throats.

Source: Authors' compilation based on data from BFN Project Sri Lanka (2009); Rajapaksha (1998); Jayaweera (1980).

3.1.7. Wild Mushrooms

Forests are the main source of food producer for people in surrounding villages. The forests produce different types of mushrooms. There are about 2500 species of mushrooms in Sri Lanka, of which only a little over 200 are known. Additionally, limited research has been conducted on edible mushrooms. However, commercially available mushrooms were first introduced to Sri Lanka in 1985, and later, the consumption of wild mushrooms declined significantly (Karunarathna et al. 2017).

Therefore, traditional knowledge is important. There are traditional beliefs about collecting and cooking different types of mushrooms because some mushrooms are poisonous. 'Mushroom hunting' became a common occurrence during this period as there is a special season that can be attributed to the growth of a particular species of fungi (Hewage 2015). Some wild mushrooms are very nutritious, tasty, and safe to eat, while others pose a serious risk to human health and can even cause death if ingested. Therefore, it is important to pick mushrooms only with someone who has extensive experience in identifying edible and poisonous mushrooms. The study found different kinds of wild mushrooms in this area. They are Urupaha, Leena Hathu, Heenvali Hathu, Idalolu, Kukul Badawel, Wea Hathu, Kotan Hathu, Kos Hathu, Uru Hathu, Piduru Hathu, Mahavali Hathu.

However, everyone must protect themselves from poisonous mushrooms. Sri Lankan people were careful with dark-colored and bad-smelling mushrooms. They avoided consuming that kind of mushroom. Also, if they found that birds and animals avoid them, they avoided eating those mushrooms.

3.2. Traditional Stable Foods and Preparation Methods

The study found that types of traditional foods, preparations, and consumption habits were more diverse than today. Sri Lankan cuisine is famous for its special combinations of herbs, spices, fish, vegetables, rice and fruits. Many of these dishes are based on rice, rice flour and coconut, and seafood plays a major role in Sri Lankan cuisine. Many Sri Lankans prefer vegetable curries and rice. Rice and curries are the staple food anywhere on the island. The curry has a huge flavor and color from the list of hot spices from Sri Lanka. These spices not only adding great flavor to the dish but also add Ayurvedic value to the dish. As Sri Lankan food culture is regionally diverse, one can expect the same food in different styles and flavors (e.g., Rice and curry Figure 1). The different dishes had unique and unique preparation systems.

(**a**) (**b**)

Figure 1. Rice and curry (**a**,**b**). Source: KavindaF (2021) (left); field work (right).

Rice preparation is a beautiful remedy for various ailments. Milk rice (Kiri Bath) is one of the specialties of Sri Lankan food culture. To prepare it, they used raw rice with coconut milk. Kiri bath plays a major role in various traditional festivals and ceremonies. Mung bean (green gram) and turmeric powders are also added to make a bowl of colorful milk rice. Yellow Rice (Kaha bath) is also one of the healthy rice preparation methods in Sri Lanka. Rice, turmeric, coconut milk and some spices mixed are tasty and healthy. Diya bath is also one of the healthy rice preparation methods in Sri Lanka. They prepared the rest of the rice mixed with water the before night. The next morning this rice mixed with coconut milk onions, salt, burned chili pieces, lime, and curry leaves were prepared for breakfast. This is a very popular food that delays hunger. Rice porridge is a traditional way of preparing healthy food. Different porridge has different health benefits. Most of used coconut milk, green leaves, garlic, and onions. "Kola Kanda" is a traditional herbal porridge made from raw rice, coconut milk, and the fresh juice of medicinally valued leafy greens. It is usually served at breakfast with a piece of jaggery (made from coconut honey or palm honey). Badi haal keda (roasted rice porridge) is also a porridge dish with roasted rice and salt. Roasted rice porridge is an energy-dense and easily digestible food for people recovering from any ailment. Ripe coconut water extract or coconut milk is rich in proteins and oils and is also an essential ingredient in Sri Lankan curries and sauce. Lightly cooked (almost boiling) coconut milk with salt, turmeric, green chilies, salt, curry leaves, and lime juice make "kiri hodi". Traditionally they had rice flour-based different types of preparation methods. Those are hoppers, string hoppers," pittu", "roti" as well as "sweets kaum", "asmi", "aluwa", "walithalapa", etc. Although some foods are only prepared for certain occasions or certain purposes, all the ingredients are natural. Some dishes are specially prepared for particular people. e.g., "Asmee", "Konda kaum".

This diet includes thin gravy "niyabalawa" mildly cooked salad "malluma" dry roasted "kabale baduma" deep fry (baduma) fry (themparaduwa). Also mixed with grated coconut or coconut oil or coconut milk and various herbs and spices are essential ingredients. Some of these supplements are paired with staple foods. For example, "lunumiris" with milk rice and coconut sambal or grated coconut with boiled potatoes or jackfruit.

They always try to eat healthy. They detected compatible foods. Incompatible foods have always been avoided. If the food had any harmful effects, it was always omitted in an ordinary meal. For example, they did not drink milk, but rather, they ate milk in its fermented form as curd. Today, scientific evidence has proven curd contains many beneficial bacteria. Cultivated and wild vegetables, especially wild green leaves and other wild plant food types were important ingredients for sauces that accompanied carbohydrate staples. The seeds were naturally hybridized and fertilized. The food was plentiful. The choice of food was dependent on the need for it. For children to overcome the burden of intestinal worms, a 'mellum' prepared from 'Eth thora' (*Cassia alata*) or 'Erabadu' (*Erythrina indica*) was used; for diabetic patients, a curry made of bitter gourd (*Mormodia aurandica*) was eaten. Similarly, there were many other dietary recommendations that could be used for therapeutic and treatment purposes. Some foods heat your body, but they use different ingredients to control that. For example, breadfruit/del (*Artocarpus altilis*) and fresh tuna are body-heating foods. To control this, eat this kind of food with coconut. Similarly, some food sources are consumed depending on the health properties of the ingredients. For example, mung beans and long beans are generally not eaten for dinner or by someone with a cold or body pain.

Also, the food that was prepared and brought to the rice field to serve the people was called "Ambula" consisting of local vegetables and rice. Another special preparation was sour fish curry (Malu Ambul thiyal), a unique spicy fish preparation with thick gamboge "Goraka" paste (Perera 2008). This shows, with diverse foodstuffs, how varieties of delicious dishes were prepared (Perera 2008). This was confirmed by Robert Knox in his book An Historical Relation of the Island Ceylon in the East-Indies. Some foods are used during a special time. "Hath maluwa" is also one of the special food recipes during the traditional new year festival. They prepared seven different food plant items that were mixed with coconut milk.

Also, "Tambum Hodi"or "Miris Hodi" is a Sri Lankan soup traditionally prepared using herbs and spices, such as black pepper, ginger, drumsticks, cinnamon, curry leaves, garlic, coriander, cumin, and fennel. "Tambum Hodi" is a special soup for various ailments including appetite and stomach ailments. It helps lower blood cholesterol, obesity, certain cancers, the immune system, and inflammatory diseases. It is also good for postpartum mothers. Therefore, the functional ingredients in this soup play a key role in digestive functions and are used as a medicine to prevent diseases.

3.3. Traditional Food Preservation

Traditionally, Sri Lankans have always kept surplus food for future use. Traditional Sri Lankan people have always kept the surplus for future use. The food preservation was due to two reasons. First, to ensure food security in the future and then to use it in difficult times. Therefore, the need for a conservation diet was satisfied during the off-season, and food waste also did not happen. Preservation was carried out using simple, appropriate, inexpensive, and sustainable indigenous technologies. The game meat is preserved in its fat, and the product is called "Kurukkal". This can only be done with high-fat meat. Therefore, Kurukkal can only be made from wild boar meat and mature venison. As well as Boiled jackfruit bulbs are half boiled, dried and preserved as "Atu kos ata" and "Atu kos madulu", respectively. During the harvest period, the seeds are collected and stored for later use. For that, they used 'Atuwa' and 'Bissa'. These play an important role in preserving traditional seeds. Until now could not find better than this structure in Sri Lanka. The study found that there were several ways to preserve it.

(1) Salting and drying: Salting and drying are combined and are popular preservative methods. These involve the use of a saline solution called dry salt or brine. When using dry salt to preserve fish, the fish is salted for two to three days before drying it in the sun and removing the excess salt. The salt concentration can be better controlled with salting. The clay pot, salt, and fish or fruit are added back to a crockpot, and then covered and left to sit for a few weeks. While absorbing the amount of water it contains, the seal adds a lot of flavors to the salt and squash, ensuring that it is not spoiled by water vapor and other contaminants in the surrounding air. Some vegetables can also be preserved by drying the sun, e.g., Atu kos (dry jackfruit), Del (breadfruit), mushroom and some vegetables.
(2) Smoking: This is probably not the easiest way to preserve meat/fish, but smoking adds extra depth to the flavor profile. This is also something that takes a long time. It takes a few more days to store and care for the food. On the other hand, smoking causes skin irritation. The traditional method is to place the meat/fish on a grill over a layer of charcoal. Usually, all traditional home kitchens had a smoking area.
(3) Immersion in honey: Traditionally, people would dip meat and fruits in honey, which was a popular way to preserve food. It acts as an antibacterial agent, and the honey protects the meat (Game meat) or fruits from oxygen exposure. In addition, it adds a complementary sweetness, enhancing its flavor.
(4) Buried under dry sand: Burying food under sand does not produce a tasty result, but with its pH level, cold temperature, and lack of light and oxygen, it creates an environment that can extend the life of some edible seeds like sand jackfruit seed (Wali kos ata).
(5) Pickling: Pickling is a combination of two preservation methods, salting and adding sugar, with a little vinegar and spices in the mix. Sri Lankans call foods treated this way pickles. Vegetable or fruit pickles are an adjunct to traditional foods. There are traditional pickles made from young papaya (*Carica papaya*), jackfruit, mango, Ceylon olive (*Elaeocarpus serratus*), and bilin (bile pigments). Traditional Sri Lankan vegetable pickles are made by mixing coconut vinegar as an acidic substance with the sharp flavor emitted from wet mustard paste and chopped drumstick bark.

3.4. Traditional Agricultural Practices

Food crops are traditional because they are accepted by rural communities as appropriate due to their customs, habits, and traditions. They believe traditional food crops have nutritional and therapeutic value. In addition, these plants are cultivated in a particular ecosystem at a specific location or are harvested in the wild or semi-wild state. Cultivation practices and methods have evolved to meet the needs of the plants as knowledge of the environmental impact of different plants cultivated by humans has improved. In ancient Sri Lanka, agriculture was indeed an organic farming system closely linked to ecosystems (Bandara 2007).

As knowledge of the environmental responses of each crop plant has improved, people have adapted cultural practices and cultivation methods to the needs of the plants. In ancient Sri Lankan people have used three main agricultural methods from the earliest times (Siriweera 1993). Namely, upland and low country paddy cultivation, Chena cultivation (transfer cultivation), and mixed home gardening methods (upcountry home gardening). They are closely associated with nature, natural ecosystems, water management, and pest control systems. Paddy cultivation is the most widely used agricultural method in Sri Lanka. This civilization has been based on paddy cultivation since ancient times. Although, there are only two main agricultural seasons in Sri Lanka called Yala and Maha seasons. The ancient agriculture system was based on both irrigated and rainfed field crops. With the construction of reservoirs and canals for irrigation more areas were cultivated. Short- and long-duration paddy varieties have been selected for cultivation in Yala and Maha seasons respectively. Paddy cultivation in the central hills is related to terrace cultivation called "Helmalu". The intensity and frequency of rainfall increases in the central highlands. In traditional rice terrace cultivation, people used to set up rice terraces using these upland slopes; so, the drainage channels for paddy cultivation had to be well managed. Conventional paddy cultivation with low rainfall in arid areas was mainly based on irrigation systems and developed from people's accumulated experience and knowledge of temperature, rainfall patterns, and soil behavior. Irrigation reservoirs are important structures that demonstrate their ability to adapt to their subsistence activities in the natural environment. These large and small irrigation systems are the best examples to illustrate the environmental protection given by sustainable agriculture. The traditional village function was based on irrigation reservoirs or streams. Rice paddy fields were located below the reservoir/canal. Due to this well-organized location, the paddy fields were gradually supplied with water from reservoirs through canals.

The second agricultural system is called shifting cultivation or slash-and-burn cultivation, which is more commonly known in Sri Lanka as "chena cultivation". Shifting cultivation is a traditional system of agriculture widely practiced in the dry zone of Sri Lanka. It involves the clearing of forest land via slashing and burning and the annual planting of crops for periods of about three years, after which the practice is moved to another forest plot, leaving the previous plot fallow. With increasing population growth and migration, the fallow period was drastically reduced until the practice was no longer viable for sustainable agriculture. In the chenas, secondary crops and vegetables were grown under rainy conditions. The secondary grains include kurakkan (millet), which was considered as the second staple, meneri (millet), thanahal (Foxtail millet), amu (Kodo millet), mustard, ginger, sesame, green gram, and black gram. Also, vegetables such as luffa, cucumber, lady's fingers (okra), snake gourd, bitter gourd, ash gourd, yellow gourd, melon, and brinjal (eggplant) were cultivated in mixed cultures in the Chenas.

The third agricultural system practiced by Sri Lankans since the earliest times is the Kandyan Forest Garden. A variety of economically valuable tree species such as spices, fruits, medicinal plants and woody species are grown. The systems are generally practiced on small family farms and in a few districts (Kandy, Matale, and Kurunegala) in the central region of Sri Lanka. Traditional root and tuber crops, yak, coconut, areca nut, vine palm (kithul), banana, sugarcane, ginger and turmeric, citrus species, and other important food and medicinal plants were commonly grown in home gardens (Siriweera 1994). In addition, it is known that the forest plays an important role in the food system of our people. It provided a wide range of food.

3.5. Food Security, Food Practice, and Diversity

When asked for their perceptions of their own consumption of traditional foods over the past 50 years, they ate "game meat" (wild meat), such as porcupine, jungle fowl, hare, wild boar, and so on. There was no shortage of any sources of "game meat" in the past, and the slaughtering of wild animals for meat (game) was allowed with restrictions. This community was consuming less of the traditional variety of foods such as healthier wild plants. This study also showed that indigenous fruits and vegetables were not so popular in this area, helping to understand the diversity of food and food transition. When decreases in consumption of a group of traditional foods were reported, the main reason given in each community was the decreased availability or unavailability of that food.

This study indicates that people do not eat wild greens, bark, roots, and mushrooms. The reasons given were not liking the foods, never having tried them, and for mushrooms uncertainty about which types were safe. However, when and if available, these conventional varieties can largely increase food security and nutrition security in this society.

Ninety percent of the households reported that their traditional foods and knowledge came from within their household or immediate family, while 10% reported that others in the community were their main source of traditional foods. Fifty percent of the households in this community reported obtaining all the traditional foods that their households wanted.

Our recent research outcome indicated that their vulnerability to food and nutrition insecurity (Weerasekara et al. 2020). In this study, we found that some endemic vegetables and fruits may be helping to achieve food security. The common vegetables and fruits in these areas have a variety of nutritional benefits, which include leafy greens consumed as part of their regular diets and used in various ways (Weerasekara et al. 2020). Since most of the households partake in small-scale agriculture, they are provided with better food security. Some of their food proportions, especially those of vegetable legumes, different kinds of mushrooms, nuts, seeds, and leafy greens from the wild, are considerably higher than in they are in urban areas. Wild foods can support households who experience financial difficulties and are important contributors to food security. Unfortunately, many did not report eating these types of wild foods. It has become customary to buy food from markets, and many are unaware of the nutritional benefits of these foods.

Our recent research study shows that wild food, such as fruits, leafy vegetables, mushrooms, tubers, and honey, increase the dietary diversity and micronutrient consumption among rural Sri Lankan communities (Weerasekara et al. 2020).

Different food items are listed in the following table (Table 6). There was a significant positive correlation between the food groups ($p < 0.01$). A total of 100% of the participants reported eating starchy staple foods, but the variety of grains, tubers, white roots, and plantation foods was wider in this area. Most of the foods that are consumed by women include rice flour products (string hoppers, hoppers, pittu, and noodles) and wheat flour products (bread, buns, koththu, and noodles). In this area, people consumed starchy staple foods, and more than 90% consumed pulses, beans, peas and lentils, nuts, and seeds. More than 50% of them consumed Vitamin A-rich fruits and vegetables, as well as other vegetables. In this study, the sample results revealed that most of the people consumed a low percentage of animal protein. There are no data about the traditional varieties.

Table 6. Type of foods consumed in this area by people.

Food Groups	Correlation	*p*-Value	Mean ± S.D.	Food Items
Starchy staple foods (grains, white roots, and tubers plantations)	-	-	1.00 ± 0.000	Rice (*Oryza sativa*), Wheat (*Triticum aestivum* L.), Jackfruit (*Artocarpus heterophyllus*), Katu ala (*Dioscorea pentaphylla*), Breadfruit (*Artocarpus*), Cassava (*Manihot esculentum*), Sweet potatoes (*Ipomoea batatas*), Kiri ala (*Xanthosoma sagittifolium*), Lotus root (*Nelumbo nucifera*), Bananas/unripe (*Musa*), Potatoes (*Solanum tuberosum* L.)
Pluses, beans peas and lentils	−0.336 **	0.000	1.08 ± 0.264	Mung bean (*Vigna radiata*), Cowpea (*Vigna unguiculata*), Long bean (*Vigna subterranean*), Bean (*Vigna angularis*), Chickpea (*Cicer arietinum*), Winged bean (*Psophocarpus tetragonolobus*), Daal, Soya (textured soy protein/TSP)
Nuts and seeds	−0.840 **	0.000	1.08 ± 0.272	Peanut (*Arachis hypogaea*), Cashew (*Anacardium occidentale*), Coconut palm (*Cocos nucifera*)
Dairy product (e.g., milk, yoghurt, and cheese	0.292 **	0.000	1.77 ± 0.457	Milk powder
Meat (all meat fish, chicken and liver or organ meat)	0.537 **	0.000	1.69 ± 0.466	Chicken, fresh or dried fish (seafood or tank fish)
Eggs	0.130 **	0.009	1.87 ± 0.337	Chicken eggs

Table 6. *Cont.*

Food Groups	Correlation	*p*-Value	Mean ± S.D.	Food Items
Dark green vegetables	0.492 **	0.000	1.37 ± 0.487	Kankung (*Ipomoea aquatica*), Gotukola (*Centella asiatica*), Mukunuvanna (*Alternanthera sessilis*), Manioc leaves (*Maniot esculenta*), Kathurumurunnga (*Sesbania grandiflora*), Pumkin leaves (*Cucurbita maxima*), Japan batu (*Sauropus androgynus*), Thebu (*Costus speciosus*), Passion fruit leaves (*Passiflora edulis*)
Fruits and vegetables	0.272 **	0.000	1.65 ± 1.132	Carrot (*Daucus carota*), Pumpkin (*Cucurbita pepo*), Papaya (*Carica papaya*), Mango (*Mangifera indica*), Radish (*Raphanus sativus*), Aubergine (*Solanum melongena*), Bitter gourd (*Momordica charantia*), Ridge gourd (*Luffa*), Snake cucumber/kekiri (*Cucumis melo*), Tomato (*Solanum lycopersicum*), Banana/Plantain flower (*Musa acuminata* Colla), Ambarella (*Spondias dulcis*), Plate brush/Thibbatu (*Solanum torvum*), Banana (*Musa paradisiaca* L.), Apple (*Malus pumila* Mill.)

Source: Table by authors. ** Statistical significance $p < 0.01$ (two-tailed), S.D.: standard deviation.

4. Discussion

These results show that many wild native edible species contribute to food security in Sri Lanka. In addition, many medicinal plants play an important role in food diversity. The traditional population of Sri Lanka diversified their diet by consuming wild, semi-cultivated, and cultivated local foods. The culture and food practices that had a major impact on household food security were the desire to control household income and to share food within the household.

Therefore, we understand the benefits of the practice of local food production for sustainable food practices. Therefore, this study understands the benefits of the practice of local food production for sustainable food practices. First of all, the closer proximity of food production and consumption can help reduce waste and energy inputs as well as recycling factories for transport, storage, and preservation (Murphy et al. 2017). Local food supply can reduce food miles, which in turn reduces carbon emissions (Cowell and Parkinson 2003; Coley 1988). Indigenous food cultures provide fresh and healthy foods because they reduce the use of preservatives and reduce their nutritional value. In short-chain production systems, fresh food is less likely to be severely processed. People's changing appetites and preference for processed foods can adversely affect their health (Provenza et al. 2015). Thus, this study shows the benefits of using native plants and developing new crops.

Obtaining new foods and local crops is one way to diversify their commercial use. Adaptation of local communities to climate change is essential to reduce poverty and food security (FAO 2016).

Native plants have adapted to their native environment. These plants require a small amount of water, fertilizer, and pest and disease control to survive, and they produce high yields (Provenza et al. 2015). Native plants can minimize soil erosion and maintain plant–microorganism–soil interactions (Balestrini et al. 2015; Hawkes et al. 2007). In addition, some research indicates that the role of below-ground interactions of plants with other organisms has been underestimated in the past (Shelef et al. 2013) and has the considerable potential to increase plant activity levels and crop yields (Drinkwater and Snapp 2007). Incorporating native food crops as temporal and spatial by-products in land management helps to maintain soil quality and prevent soil degradation. Similarly, intercropping helps maintain the soil quality and improves nitrogen uptake (Eaglesham et al. 1981), reduces the number of weeds (Liebman and Dyck 1993), and provides farmers with a higher net income (Yildirim and Guvenc 2005). The use of indigenous species can have a positive impact on human health. With the introduction of specially processed foods, the so-called Western diet has changed the main nutritional properties of human food. In addition, the food industry has opted for fewer fruits and vegetables, preferring varieties that are less rich in phytochemicals than their traditional counterparts are (Robinson 2013). The diversification and expansion of the use of local plants, coupled with the cultural practices of preparing these foods, adds health-promoting plants to people's diet and eliminates the obvious economic costs of such practices (Provenza et al. 2015). According to our study, indigenous plants used by humans for centuries are highly nutritious, tasty, and easily digested foods. Therefore, the use of native plants as a supplement in food production has considerable advantages and, due to their richness in plant substances, they improve both human health and food security (Provenza et al. 2015). The selection of native plants not only increases food diversity for humans, but also diversifies agricultural entrepreneurship and preserves genetic diversity to improve the environmental conditions.

However, it is well known that both natural and human factors contribute to the erosion of plant diversity. In the evolutionary process, plants die out or evolve into new species, but the extinction rate due to human factors is much higher. Political, economic, and social factors have a direct influence on genetic erosion, especially the erosion of food crops.

Traditional Sri Lankan's subsistence agriculture, farming methods, food system and food culture were challenged during the British era of the 19th century. They promoted the cultivation of tea, coffee, and rubber on plantations. Plantation agriculture damaged traditional agriculture and plant diversity. For a long time, the British did not promote traditional agriculture, hoping to break the backbone of traditional agriculture. This had a profound impact on traditional food varieties and farming practices. European food habits have also had a serious impact on traditional Sri Lankan food and farming methods (Weerasekara et al. 2018). Although the introduction of these plants can enrich the plant diversity of Sri Lanka the damage to the diversity of traditional food crops is greater as various plants are grown on a large scale to meet the high demand for exotic varieties. Some of the policies and strategies of the governments that took power before and after independence also led to the decline of traditional food crops. The motto of the green revolution initiated in the 1960s was to increase productivity. The existing traditional varieties were deemed to be unsuitable for the new conditions, and a few new varieties were bred and cultivated. The monoculture is the accepted cultivation method in modern agriculture.

Thousands of food crops grown in farmers' fields are seriously threatened by the cultivation of certain varieties of commercial or useful crops. Deforestation also has profound effects on the genetic erosion of plants and the food security of the people in this region. About 200,000 square kilometers of forests disappear every year.

The diversity of wild edible plant varieties and edible plants is seriously threatened. For example, cereals such as millet, which were a major contributor to carbohydrate intake at the start of this century, are hardly grown anymore. This situation has been aggravated by commercial agriculture. The relative advantage of commercial agriculture has led to the cultivation of a multitude of plants in farmers' fields to meet food needs, displacing several commercially important plants. In modern agriculture, much of the food is destroyed as "weeds". Consider the reasons why a wide range of traditional food plants are important for human survival. In Sri Lanka, the diversity of edible wild plants is seriously threatened. For example, millet is hardly grown anywhere due to commercial crops.

Food security depends on the diversity of food crops. It is now widely recognized that increasing the number of food plant varieties can bring benefits to the community both in the short and long term; so, the search for new alternatives or unconventional plant resources is desperately needed to diversify current agriculture. In this context, underused traditional foods are of particular importance.

Despite the improvement of many public health indicators, malnutrition remains a problem in Sri Lanka, particularly among women and children. Diseases and a lack of high-quality food are two major causes of malnutrition. These problems include poor infant growth, infants with a low birth weight (LBW), poor maternal nutrition, and micronutrient deficiencies. The preparation and conservation of local seasonal foods are important to improve the family's nutritional status. Also, traditional local plants can be effectively used as a low-cost food source for low-income groups. In this context, traditional food crops are of great benefit as they make an important contribution to meeting the nutritional needs of the general, popular and even the rural poor. These food crops adapt well to unfavorable environmental conditions, and some of them can even be grown in marginal soils that are resistant to pests and diseases. Therefore, they require less attention and fewer resources. They are nutritious and have the same value as socially accepted foods and can serve as a staple or as a supplement. Since most of them are harvested from the environment or use fewer agrochemicals than commercial crops do, they can be considered as fresh and healthy foods. Consuming a variety of plant species ensures that the human body is getting all the nutrients it needs.

They provide seasonal food and give additional income to the farmers. Many plants are used as medicine, and firewood, and can be a useful household appliance. Traditional indigenous knowledge for sustainable agriculture can be directly promoted as an important crop such as traditional yam and green leaves in the future incorporating traditional food plants into agriculture development programs. This plant has beneficial properties in terms of nutrition and value, resistance to adverse soil conditions, resistance to drought and resistance, and benefits to disease so that it can be used as genetic material or future crop improvement programs. It contains invaluable local knowledge on the cultivation, preparation, nutrition, and storage of traditional food crops and can be incorporated into future agricultural programs with or without modifications. Traditional farmers had extensive and unique knowledge of plants and their agricultural practices.

The traditional Sri Lankan people knew about the ecosystem in which they lived and applied methods to ensure the sustainable use of natural resources. They not only used a wide range of plant species, but also developed a large number of varieties adapted to different climatic conditions. They knew how to prepare tasty and nutritious food and how to safely store leftovers for later consumption. Cookery is one of the sixty-four noble arts of ancient Sri Lanka.

People ate different foods. Rice was the staple food and was prepared in many ways. They knew that preparing certain foods could destroy their toxic components. They also realized that certain foods are not allowed to be eaten together because they can be harmful to humans. They tried to prepare good food for different ages and conditions. This extensive knowledge of preparation and nutrition is especially integrated with women. This knowledge was transmitted from mother to daughter from generation to generation. Much of this knowledge is rapidly disappearing, but knowledge of the traditional Ayurvedic system and nutrition is still present. This knowledge has been transmitted orally and practically from one generation to another. Unfortunately, this knowledge and the skills related to traditional agriculture and food culture are rapidly disappearing. According to this sustainable agricultural policy, attention should be paid to traditional food cultures and local knowledge.

Author Contributions: All authors contributed equally to this research work.

Acknowledgments: This research was facilitated by the Department of Organic Food Quality and Food Culture at the University of Kassel, Germany. The authors express their thanks to all the interviewees in the field who dedicated their time and allowed the authors to participate in their meetings. Without their unconditional support, it would have been impossible to complete this field research study. Last, but not least, we would like to thank the anonymous reviewers of the MDPI Books for their critical and constructive comments.

Conflicts of Interest: The authors declare no conflict of interest.

References

Balestrini, Raffaella, Erica Lumini, Roberto Borriello, and Valeria Bianciotto. 2015. Plant-soil biota interactions. *Soil Microbiology, Ecology and Biochemistry* 2015: 311–38.

Bandara, Sarath. 2007. *Nature Farming-Integration of Traditional Knowledge Systems with Modern Farming in Rice*. Leusden: ETC/Compas.

BFN Project Sri Lanka. 2009. *The Biodiversity for Food and Nutrition Project*. Peradeniya: National Agriculture Information & Communication Centre—Department of Agriculture.

Biodiversity for Food and Nutrition. 2018. The Biodiversity for Food and Nutrition Project. Available online: http://www.b4fn.org/resources/species-database/ (accessed on 2 July 2023).

Boyer, Jefferson. 2010. Food security, food sovereignty, and local challenges for transnational agrarian movements: The Honduras case. *The Journal of Peasant Studies* 37: 319–51. [CrossRef]

Chappell, Michael Jahi, and Liliana A. LaValle. 2011. Food security and biodiversity: Can we have both? An agroecological analysis. *Agriculture and Human Values* 28: 3–26. [CrossRef]

Coley, P.D. 1988. Effects of plant growth rate and leaf lifetime on the amount and type of anti-herbivore defense. *Oecologia* 74: 531–36. [CrossRef] [PubMed]

Cowell, Sarah J., and Stuart Parkinson. 2003. Localisation of UK food production: An analysis using land area and energy as indicators. *Agriculture, Ecosystems & Environment* 94: 221–36.

De Fonseka, R.N., and S. Vinasithamby. 1971. *A Provisional Index to the Local Names of the Flowering Plants of Ceylon*. Peradeniya: Department of Botany, University of Ceylon.

De Soucey, Michaela. 2010. Gastronationalism: Food traditions and authenticity politics in the European Union. *American Sociological Review* 75: 432–55. [CrossRef]

Drinkwater, Laurie E., and Sieglinde S. Snapp. 2007. Understanding and managing the rhizosphere in agroecosystems. In *The Rhizosphere*. Cambridge: Academic Press, pp. 127–53.

Eaglesham, A.R.J., A. Ayanaba, V. Ranga Rao, and D.L. Eskew. 1981. Improving the nitrogen nutrition of maize by intercropping with cowpea. *Soil Biology and Biochemistry* 13: 169–71. [CrossRef]

FAO. 2016. The State of Food and Agriculture 2016 (SOFA): Climate Change, Agriculture and Food Security. Available online: http://www.fao.org/publications/card/en/c/18679629-67bd-4030-818c-35b206d03f34/ (accessed on 13 May 2020).

Godfray, H. Charles J., and Tara Garnett. 2014. Food security and sustainable intensification. *Philosophical Transactions of the Royal Society B: Biological Sciences* 369: 20120273. [CrossRef]

Godfray, H. Charles J., John R. Beddington, Ian R. Crute, Lawrence Haddad, David Lawrence, James F. Muir, Jules Pretty, Sherman Robinson, Sandy M. Thomas, and Camilla Toulmin. 2010. Food security: The challenge of feeding 9 billion people. *Science* 327: 812–18. [CrossRef]

Hammer, Karl, Nancy Arrowsmith, and Thomas Gladis. 2003. Agrobiodiversity with emphasis on plant genetic resources. *Naturwissenschaften* 90: 241–50. [CrossRef]

Hawkes, Christine V., Kristen M. DeAngelis, and Mary K. Firestone. 2007. Root interactions with soil microbial communities and processes. In *The Rhizosphere*. Cambridge: Academic Press, pp. 1–29.

Helvetas Sri Lanka. 2001. *Sustainable Farming System*. Sri Lanka: Helvetas Sri Lanka.

Hewage, Dammika. 2015. Traditional Knowledge of Edible Wild Mushrooms in a Village Adjacent to the Sinharāja Forest. *Journal of the Royal Asiatic Society of Sri Lanka* 2015: 77–95.

Jayaweera, DMA. 1980. *Medicinal Plants (Indigenous and Exotic) Used in Ceylon*. Colombo: National Science Council of Sri Lanka, Part II.

Karunarathna, Samantha C., Peter E. Mortimer, Jianchu Xu, and Kevin D. Hyde. 2017. Overview of research of mushrooms in Sri Lanka. *Revista Fitotecnia Mexicana* 40: 399–403. [CrossRef]

KavindaF. 2021. Rice, Food, Dish Image. Available online: https://pixabay.com/photos/rice-food-dish-meal-curry-6324799/ (accessed on 1 August 2023).

Knox, Robert. 1983. *An Historical Relation of the Island Ceylon in the East-Indies*. Colombo: K.V.G. de Silva and Sons (Co.) Ltd.

Liebman, Matt, and Elizabeth Dyck. 1993. Crop rotation and intercropping strategies for weed management. *Ecological Applications* 3: 92–122. [CrossRef] [PubMed]

Massy, Charles. 2020. *Call of the Reed Warbler: A New Agriculture—A New Earth*. Brisbane: University of Queensland Press.

Murphy, Brian, Paul Crosson, Alan K. Kelly, and Robert Prendiville. 2017. An economic and greenhouse gas emissions evaluation of pasture-based dairy calf-to-beef production systems. *Agricultural Systems* 154: 124–32. [CrossRef]

Ofuya, ZM., and V Akhidue. 2005. The role of pulses in human nutrition: A review. *Journal of Applied Sciences and Environmental Management* 9: 99–104. [CrossRef]

Perera, ANF. 2008. Sri Lankan Traditional Food Cultures, and Food Security. *Economic Review* 2008: 40–43.

Provenza, Frederick D., Michel Meuret, and Pablo Gregorini. 2015. Our landscapes, our livestock, ourselves: Restoring broken linkages among plants, herbivores, and humans with diets that nourish and satiate. *Appetite* 95: 500–19. [CrossRef]

Rajapaksha, Udaya. 1998. *Traditional food plants in Sri Lanka*. Colombo: Hector Kobbekaduwa Agrarian Reserarch and Training Institute.

Ritchie, Donald A. 2014. *Doing Oral History*. Oxford: Oxford University Press.

Robinson, Jo. 2013. Breeding the nutrition out of our food. *New York Times*. May 25. Available online: https://www.nytimes.com/2013/05/26/opinion/sunday/breeding-the-nutrition-out-of-our-food.html (accessed on 7 March 2022).

Sethi, Simran, and Therese Plummer. 2015. *Bread, Wine, Chocolate*. New York: Harpercollins Publishers Incorporated.

Shelef, Oren, Peter J. Weisberg, and Frederick D. Provenza. 2017. The value of native plants and local production in an era of global agriculture. *Frontiers in Plant Science* 8: 2069. [CrossRef]

Shelef, Oren, Yael Helman, Ariel-Leib-Leonid Friedman, Adi Behar, and Shimon Rachmilevitch. 2013. Tri-party underground symbiosis between a weevil, bacteria and a desert plant. *PLoS ONE* 8: e76588. [CrossRef]

Sherasia, P. L., Manget Ram Garg, and B. M. Bhanderi. 2018. *Pulses and Their By-Products as Animal Feed*. New York: United Nations.

Shiferaw, Bekele, Boddupalli M. Prasanna, Jonathan Hellin, and Marianne Bänziger. 2011. Crops that feed the world 6. Past successes and future challenges to the role played by maize in global food security. *Food Security* 3: 307–27. [CrossRef]

Siriweera, Indrakerthi. 1993. *Sri Lankawe Krushi Ithihasaya (till 1500 AD)*. Colombo: S. Godage and Brothers, pp. 9–29.

Siriweera, W. I. 1994. *A Study of the Economic History of pre-Modern Sri Lanka*. Uttar Pradesh: Vikas Publishing House Private.

Swami, Shrikant Baslingappa, N. J. Thakor, P. M. Haldankar, and S. B. Kalse. 2012. Jackfruit and its many functional components as related to human health: A review. *Comprehensive Reviews in Food Science and Food Safety* 11: 565–76. [CrossRef]

Tennent, James Emerson. 1860. *Ceylon: An Account of the Island, Physical, Historical, and Topographical with Notices of Its Natural History*. London: Longman, Green, Longman, and Roberts, vol. 2.

Thompson, Paul. 2017. *The Voice of the Past: Oral History*. Oxford: Oxford University Press.

Weerakkody, Nimsha S., Nola Caffin, Mark S. Turner, and Gary A. Dykes. 2010. In vitro antimicrobial activity of less-utilized spice and herb extracts against selected food-borne bacteria. *Food Control* 21: 1408–14. [CrossRef]

Weerasekara, Permani C., Chandana R. Withanachchi, G. A. S. Ginigaddara, and Angelika Ploeger. 2018. Nutrition transition and traditional food cultural changes in Sri Lanka during colonization and post-colonization. *Foods* 7: 111. [CrossRef]

Weerasekara, Permani Chandika. 2013. The Impact of Policy Responses to the Food Price Crisis and Rural Food Security in Sri Lanka. *Future of Food: Journal on Food, Agriculture and Society* 1: 61–70.

Weerasekara, Permani C., Chandana R. Withanachchi, G.A.S. Ginigaddara, and Angelika Ploeger. 2020. Understanding Dietary Diversity, Dietary Practices and Changes in Food Patterns in Marginalised Societies in Sri Lanka. *Foods* 9: 1659. [CrossRef]

Yildirim, Ertan, and Ismail Guvenc. 2005. Intercropping based on cauliflower: More productive, profitable and highly sustainable. *European Journal of Agronomy* 22: 11–18. [CrossRef]

Eliminating Hunger: Yam for Improved Income and Food Security in West Africa

Beatrice Aighewi, Norbert Maroya, Robert Asiedu, Djana Mignouna, Morufat Balogun and P. Lava Kumar

1. Introduction

Yam is a monocotyledonous tuber crop in the family Dioscoreaceae. The genus includes more than 600 species worldwide (Mondo et al. 2020), but only a few are cultivated. *Dioscorea rotundata* is the most valuable and widely cultivated species in West Africa, followed by *D. alata. D. cayenensis, D. dumetorum, and D. esculenta*, which are also cultivated in limited quantities in the region. Yam is a crop of great value in many communities in West Africa, and it is mainly cultivated for its underground tubers, although some species also produce aerial tubers. The crop has the highest value of the aggregate production compared to other crops in West Africa (Elbehri 2013; Hollinger and Staatz 2015), with an apparent per capita consumption that increased fastest in major producing countries of Ghana and Nigeria over the period 1980–82 to 2007–09 (Hollinger and Staatz 2015). While some yam species are important as food, others provide useful pharmaceutical products (Price et al. 2016; Tohda et al. 2017). Diosgenin and dioscorin are compounds isolated from yam and used in the pharmaceutical industry (Obidiegwu et al. 2020). However, some yam species are not edible due to poisonous substances (Dave et al. 2020; Joob and Wiwanitkit 2014; Kyung-Sik and Taek Heo 2015; Yoon et al. 2019).

Yam is grown as an annual with a duration from planting to harvest ranging from 8 to 12 months, depending on the agroecological conditions. Varieties in the forest region typically have a longer crop duration due to the longer rainy season than crops in the savanna. About 74.9 million tons of yam tubers are produced annually in the world on about 8.9 million hectares of land with an average yield of 8.5 t/ha (FAOSTAT 2021). Africa contributes 97.8% to world production, and Benin, Côte d'Ivoire, Ghana, Nigeria, and Togo account for 93.9% of world production (FAOSTAT 2021). Nigeria alone produces 66.9% of global production. Yam is the fourth most utilized root and tuber crop globally after potato (*Solanum* spp.), cassava (*Manihot esculenta*), and sweetpotato (*Ipomoea* spp.), and it is the second in West Africa after cassava (Lev and Shriver 1998).

Where yam is produced, it is a significant food source and cash crop, thus combating malnutrition, food insecurity, and poverty (Obidiegwu et al. 2020). In Nigeria and Ghana, 31.8% and 26.2% of the population depend on yam for food and income security. Yam tubers have better nutritional attributes than other root crops (Shajeela et al. 2011). The tubers are a good source of essential dietary supplements such as protein and well balanced essential amino acids (Baah et al. 2009). Even where yam is not a significant crop in some parts of Africa, it is considered a famine food for small and marginal rural families and forest-dwelling communities during periods of food scarcity (Ngo et al. 2015). In West Africa, yam is most preferred when processed fresh. It is mainly consumed boiled in sauce or soup (using pieces directly from the tuber or frozen chips from a supermarket), made into a dough by pounding boiled tuber pieces or reconstituting from flour (as instant pounded yam or *amala*), roasted, or fried.

Yam is often referred to as the "king of crops" in West Africa, and it is essential in the socio-cultural life of the population. It is used in ceremonies related to fertility and marriages, cultural rites of passage, thanksgiving, petitions, and annual festivals held to celebrate its harvest (Obidiegwu and Akpabio 2017; Nweke 2016). Poor seed and ware yam storage systems cause seasonality in the price and quality of available yams. Surplus during peak production season leads to price 'crashes', affecting farmers, while scarcity at other times causes a lack of affordability by consumers and undernutrition. Cropping of yam, especially seed, throughout the year therefore contributes immensely to the regional requirement for food and income.

2. Trends and Systems of Yam Production

A 70 percent increase in global food production is required to feed an additional 2.3 billion people by 2050 (FAOSTAT 2021), while food production from developing countries should almost double. Yam production in West Africa increased from 8.3 million tonnes in 1961 to 74.2 million tonnes in 2019. Figure 1 shows a decline in the average world yield to a present average of 8.5 t/ha, although yam has a potential of 40 to 50 t/ha (Adewumi et al. 2022; FAOSTAT 2021). The increase in production is due to an expansion of the area cultivated (Frossard et al. 2017). However, low productivity in most production systems results from poor quality of soils and inputs, accompanied by little or no application of improved agronomic practices (limited weeding, low planting density, persistent virus infections, no pest and disease management). Although there are improved and released varieties in major producing countries such as Nigeria and Ghana, local landraces still dominate farmers' varieties mainly due to challenges associated with large-scale multiplication of seed to meet demand. Smallholder farmers who constitute the most producers mainly cultivate yam in the richest soils available to them. As fertility reduces, they use the land for other crops while yam cultivation is moved to more fertile soils. However, with an increase in population and pressure on agricultural land, the availability of fertile land is reducing. Therefore, marginal soils and non-traditional yam lands are put under cultivation to obtain the same yields or less.

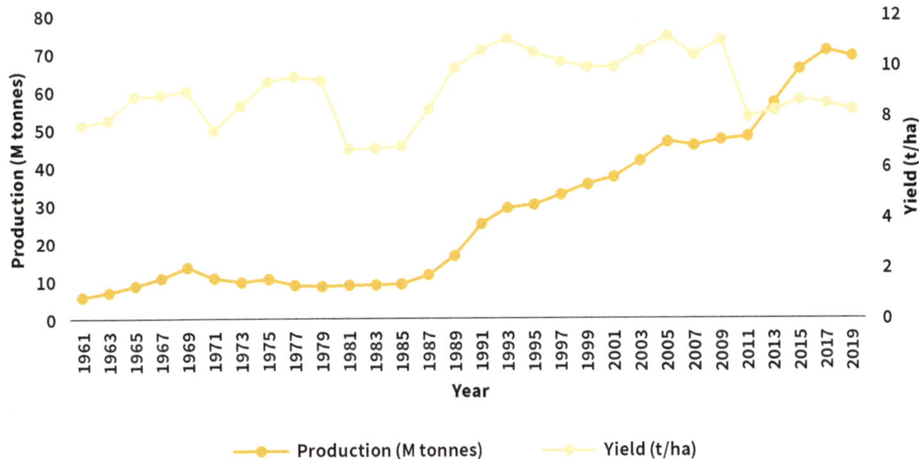

Figure 1. Trends in yam production and yields in West Africa, 1961–2019. Source: (FAOSTAT 2021).

Traditionally, farmers use whole tubers or large setts (sliced portions of tubers) of 200 to 500 g or more as planting materials (seeds) and plant in mounds prepared using topsoil. Most yam farmers grow the crop without inputs such as fertilizer or undertake crop management practices to control pests and diseases (Mignouna et al. 2015). It is frequently intercropped, but sole crops are also common in West Africa. At harvest, farmers sort small- to medium-sized tubers (200 g to 1 kg) for reuse as seeds, while large-sized tubers (>1 kg) are used as food. They are stored and eaten piecemeal or sold in markets when cash is needed. The seed-sized tubers are stored to break dormancy before planting in the next cropping season, or they are planted in newly prepared fields where they will be dormant until the start of the rainy season three to four months later.

3. Opportunities for Increased Income, Employment, and Food Supply in Seed Yam Production

Seed tubers are expensive, sometimes accounting for 63% of total variable production costs, and are bulky and expensive to transport (Manyong 2000; Agbaje et al. 2005). As planting material, the yam tuber has a meager multiplication ratio of 1:3 to 1:5 compared to some cereals (1:300). The low multiplication rate is a critical constraint to increasing yam production and productivity, resulting in the scarcity and high cost of quality seed yam (Aighewi et al. 2017). The consequence is that improved, released yam varieties are rarely found with farmers, and the use of farmer-saved seed is the norm. About a third of harvested yam tubers are reserved for seed for the next crop. Thus, from a West African production of 70.8 million tonnes, it is estimated that over 23.6 million tonnes, some of which would have been used as food, is reserved for planting the next crop (Figure 2). Continuous recycling of planting material often reduces quality by accumulating pests and diseases. After planting, some farmers can still reserve up to a third of the quantity of seed planted to replace those that would eventually not sprout due to poor quality (Aighewi 1998), further depleting a potential food and income source.

Figure 2. Yam tubers of up to 1 kg, which are good for local food and export, are used as seed and cut into minisetts of 30 to 50 g for multiplication. Source: Pictures by authors.

The traditional methods to produce seed yam include cutting ware-sized tubers into seed size, harvesting the same crop twice (the first harvest is used for food and the second for seed), sorting seed size tubers, and using only the head portions of tubers (Aighewi et al. 2015). These options multiply yam very slowly with no guarantee of seed quality. This situation formed the basis for the Yam Improvement for Income and Food Security in West Africa (YIIFSWA) initiative of the International

Institute of Tropical Agriculture (IITA). The focus was on developing methods to improve the quality of seed yam and rapidly multiply the seed to meet the needs of farmers for increased productivity. Any technology to rapidly multiply the crop is an opportunity to develop seed yam production-related businesses as a source of income, especially for youths and women. The development of hitherto non-existent seed yam enterprises will fill the gap in the supply of quality seed yam and create job opportunities.

4. The Key Elements of the YIIFSWA Project

YIIFSWA is a ten-year project, implemented from 2011 to 2021 with grants provided by the Bill and Melinda Gates Foundation (BMGF) to the International Institute of Tropical Agriculture (IITA) in two phases of five years each. The project's focus was to improve the productivity of yam through enhancing its seed systems.

4.1. Summary of Achievement of the First Phase of the YIIFSWA Project

The first five years of the YIIFSWA project (September 2011 to December 2016) facilitated activities to increase yam productivity (yield and net output) by 40% for 200,000 smallholder farmers in Ghana and Nigeria. Another goal was to generate international research goods that will double the income of 3 million yam producers in a 10-year horizon. During this first phase, YIIFSWA initiated the development of formal yam seed production systems, leading to the first set of certified seed yams for sale to ware yam producers in May 2016. This significant step in developing the yam seed market was critical to sustaining commercial production and marketing of high-quality seed yam. Diagnostics tools for virus detection and technologies for elimination of virus infected sources from seed production, high-ratio propagation technologies (HRPTs) such as the Temporary Immersion Bioreactor System (TIBS) and aeroponics, and seed yam quality management protocols as well as quality standards for seed yam certification were developed.

Over 65,000 farmers were trained to improve their seed production techniques using the adaptive yam minisett technique (YIIFSWA 2017), which is similar to what is used in the traditional seed production system (Aighewi et al. 2014). Yam-growing households in Ghana and Nigeria were characterized through a baseline survey (Mignouna et al. 2014a, 2014b). Economic assessments showed gains from the various technologies emanating from YIIFSWA and their relative profitability (Mignouna et al. 2020). Thousands of training materials (flyers, books, videos, and posters) were produced and disseminated to stakeholders to improve their capacities in seed and ware yam production, handling, and marketing (YIIFSWA 2017).

5. Novel Yam High Ratio Propagation Systems: Outcomes and Prospects in Adaptation to Climate Change and Nutrition Security

The YIIFSWA project developed and standardized a tissue-culture-based heat therapy combined with a meristem culture procedure to generate virus-free yam planting material. This procedure had a 73% success rate in eliminating the yam mosaic virus (YMV), the most frequently contaminating virus in yams in West Africa (Balogun et al. 2017a). This procedure established YMV-free stocks of 25 yam genotypes consisting of improved varieties and landraces, which were used as stocks for rapid bulking of clean planting material using the TIBS, an advanced high-ratio in vitro propagation technology also standardized by YIIFSWA.

Plantlets from TIBS are of better quality than those from conventional tissue culture (CTC). TIBS plantlets are more vigorous and resilient to post-flask acclimatization. Due to more efficient process control, large batches are handled more easily for scale-up propagation with lower risks of mix-ups. The propagation ratio in TIBS was five to six per plantlet every eight to ten weeks compared to three to four every 12 to 16 weeks in CTC (Balogun et al. 2017b). Up to 300 single-node vine cuttings were obtained per plant that was derived from TIBS plantlet and grown aeroponically (Maroya et al. 2014c, 2017), while the drip system hydroponics further saves on electricity needs in the production system (Balogun et al. 2020, 2021). About 100 single-node vine cuttings can be made per TIBS plantlet using hydroponics after eight weeks of growth, with 95% rooting success followed by field planting. After five months of hydroponic growth, the seed tubers ranged from 5 g to 220 g per plant.

Aeroponics yam propagation started under the YIIFSWA project in 2013 at IITA-Ibadan, Nigeria (Maroya et al. 2014a, 2014c). Atomizing nozzles ensure the most effective nutrient delivery in this system by turning the nutrient solution into a mist, which is absorbed through the cell walls of the plant's roots by osmosis. Genotypes of both *D. rotundata* and *D. alata* were successfully propagated with aeroponics using both pre-rooted and freshly cut two-node vine cuttings. Three types of planting materials are generated from the aeroponics production system: mini-tubers which range from 0.2 g to 110 g depending on the genotype, harvest age, and the composition of the nutrient solution, the aerial bulbils from both water yam and white yam varieties, and single-node cuttings from vines (Maroya et al. 2014b). All the propagules had a propagation rate of over 90%. Yam vine cuttings were previously known to grow slowly, but those from aeroponic plants produced new leaves 14 to 21 days after planting. The single-node vine cuttings from aeroponic plants produce an average of 200 to 300 cuttings per plant in four to six months. A manual was produced to help private seed companies to establish their aeroponics systems with step-by-step instructions for building and managing the system for foundation seed production (Maroya et al. 2017). In 2017, the YIIFSWA-II project funded each of five selected seed companies with USD 30,000, half of the cost to build a screen-house and an aeroponics system. The project produced and distributed virus-indexed plantlets for multiplication to seed companies. Relevant stakeholders were trained, including postgraduate degree students, to apply various propagation techniques through participatory research. After training, new technical information on the high ratio propagation of yam was constantly provided, accompanied by backstopping partners and other stakeholders.

The capacities of the seed regulatory agencies of Nigeria, the National Agricultural Seed Council (NASC); and of Ghana, the Plant Protection and Regulatory Services Directorate (PPRSD); were also enhanced. As coordinators in the management of seed quality in their respective countries, they were trained and provided with equipment to ensure that high-quality seeds of authentic varieties were produced and sold.

6. Improved Yam Seed Systems as Key to Building Resilience in Production to Enhance and Sustain Food Security

A functional yam seed system is essential to build resilience in the ware yam production systems. As the most critical and expensive input of yam production, the seed should be healthy to generate healthy plants in the field and reduce yam's current high postharvest losses estimated at 30%. Seed health influences the shelf life of harvested ware tubers or postharvest storage quality since pests and diseases are easily transferred from the field into storage and over different seasons through the seed. Therefore, a yam seed system must have protocols to ensure the production of seed tubers at a high rate of multiplication. Rapid propagation of pest- and disease-free seeds and a network of commercial seed producers within a formal seed system are crucial for disseminating improved varieties for sustained and increased productivity. Extension services were used to educate farmers on plant health management in the field to recognize pest and disease symptoms and carry out positive selection practices to sustain good yields.

Technologies for breeder seed production were transferred to the following NARIs, the National Center for Genetic Resources and Biotechnology (NACGRAB) and the National Root Crops Research Institute (NRCRI) in Nigeria, and the Crops Research Institute (CRI) and Savanna Agricultural Research Institute (SARI) in Ghana. These institutions were provided with equipment and supplies for backup electricity, TIBS, post-flask handling and documentation, and initial clean stock of planting materials. The NARIs now provide clean breeder planting materials to private seed companies for foundation seed production.

The innovations developed by YIIFSWA to improve yam seed systems have added value to the yam crop by using vines for propagation. Before now, yam vines were mainly considered as the apparatus for manufacturing photosynthates for storage in tubers. In the new seed production systems, leafless nodal cuttings are used in TIBS, while those with leaves are planted in various hydroponic systems or rooted and planted in the field (Figure 3). The nodal cuttings with leaf can also be planted directly in the field for seed yam production. Methods of producing minitubers of less than 10 g, which perform excellently in seed yam production (Aighewi et al. 2021), have also been developed. If widely adopted by seed companies, there would be less pressure on using yam tubers for food and seed. Six seed companies in Nigeria and three in Ghana are using the innovations developed and scaled by the YIIFSWA project for seed production.

Figure 3. Types of yam planting materials (**a**). In vitro single node cuttings for multiplication in temporary immersion bioreactor systems; (**b**) Ex vitro nodal cuttings with leaf for multiplication in various hydroponics systems, soil, and composite media; (**c**). Minitubers from ex vitro single node plants for vine and tuber planting material production in the field or screenhouse. Source: Pictures by authors.

The HRPTs developed by YIIFSWA are contributing to the efficient production of end-user preferred improved varieties. According to previous studies by CIRAD and partners, flour from dried chips made from the late maturing, multiple tubering (5–10 tubers per plant) varieties of *D. cayenensis* and *D. rotundata* with relatively high dry matter content (around 40%) and tolerance to poor soil conditions were most demanded for preparation of *amala,* a popular food in Nigeria and Benin. This link between food demand and preferred variety confirms that seed systems need to choose their promoted varieties based on market demand. However, the delivery rate of improved varieties is slow due to the low propagation ratio in traditional yam cultivation, where cropping is seasonal, with a single annual production cycle that is significantly limited by season (Orkwor and Asadu 1998). Weather elements are also crucial for yam production ranging from water, light, nutrients, and temperature. Yam requires 20 °C to 30 °C for optimum growth. However, according to the IITA's Geospatial laboratory, in 2015, temperatures lower than 20 °C were recorded only in January and December, a period that is typically outside the crop growth period, while at least 7 out of the 12 months in 2005, 2010, 2015 and 2017 recorded higher than 30 °C (Personal communication) with associated rising air temperature and carbon dioxide levels. The impact of climate change on the yam yield, including its vulnerability to climate-change-related soil conditions, has been discussed by Srivastava et al. (2012).

Just as drought causes significant losses, floods also cause significant yield losses (Balogun and Gueye 2013). These climate-related scenarios force farmers to vacate flood- or drought-prone farmlands or risk immense losses due to inadequate photosynthetic rates, tuber bulking, and premature senescence. While there have been significant breakthroughs in the real-time availability of geographical information systems, its utilization in timing yam production is yet to be entirely scaled and adopted at the level of producers and processors. The scarcity of clean planting materials (Aighewi et al. 2015, 2021), dormancy of the tuber for about four months, and uncontrolled sprouting after dormancy break limits the control of the cropping calendar by farmers and reduces profits (Craufurd et al. 2001). This established need for resilient, climate-smart mitigation strategies is also addressed by the gains from the YIIFSWA project, especially in terms of yam cropping cycles. Seed production can be carried out in controlled environments during the offseason, and ware yam production timed to maximize the available rainy period.

Table 1 shows the timing of seed yam production based on the traditional system relative to the HRPTs. Two cropping cycles are possible per year, from May to December and October to August. Previously, 14 t/ha at 40,000 stands for seed yam production was recorded. If the same land is used, up to 30 t/ha will be produced per cycle when vine cuttings are used, giving 60 t/Ha per year. With continuous availability of seed yam from the HRPTs, ware yam can be produced throughout the year, combining the traditional cycle of October to July with the HRPT cycle of May to December. This scheme is more amenable to delayed rains if supplemental irrigation is provided. Significantly more land can also be used for yam production due to 500-fold more seed yam, which culminates in more food availability.

Table 1. Seed yam production based on novel high ratio propagation technologies.

	Before	Current Revolution (Now)				Prerequisite
Stock	One tuber	One tuber				
Cut portions	6 gives 6 mother plants	6 gives 6 mother plants				
	October to August	January to May/June		July to December		
Year 1	6 mother plants	6 mother plants × 30 nodes	180 Tissue Culture plantlets in laboratory cleaned			CTC Lab
Year 2	6 cut portions per mother plant = 36	180 × 3 = 540 plantlets in 11 TIBS	540 × 3 = 1620 plantlets in 32 TIBS	1620 × 2 = 3240 seedlings in Vivipak	Production of 3240 breeder plants in Aeroponics and Hydroponics	TIBS Lab, AS, HS, Air Cooling
Year 3	6 cut portions from 36 mother plants = 216	Production of 3240 × 100 = 324,000 single node vine seedlings from AS/HS plants in nurseries		Field planting of breeder vine seedlings for foundation seed, 324,000 seed tuber production		Supp irrigation
Year 4	6 cut portions from 216 mother plants = 1296	Dormancy of foundation seed tubers		Field planting of foundation seed tubers certified seed production using AYMT (324,000 × 2 = 648,000)		Supp irrigation

AS = aeroponic system; HS = hydroponic system. Source: Maroya et al. (2022).

van Etten et al. (2017) has established that the access farmers have to quality seed and the functionality of seed systems in relation to production, distribution, innovation, and regulation determines the efficacy in contributing to sustainable agrobiodiversity and food systems. Available evidence has shown that the YIIFSWA II project has impacted all four aspects. However, it is necessary to incorporate good practices for efficient soil and water management as well as good agronomic and agroecological practices that mitigate and enhance the adaptation to climate change in future projects.

An additional benefit of the novel high-ratio propagation system is the mitigation against hidden hunger. Nutrients can be dosed to yam grown in aeroponics and hydroponics systems. High-quality planting materials of improved varieties are disseminated as reported by a study conducted by YIIFSWA (unpublished) to determine the nutrient components that limit the propagation ratio and quality of plantlets in the TIBS. Using white and water yam varieties, Kpamyo and Swaswa, respectively, the nitrogen (N), potassium (K), and phosphorus (P) concentrations in culture medium were determined every two weeks for ten weeks and in plantlets at transplanting. After eight weeks of culture, the pH and refractometer values for medium acidity and sugar concentration were determined in the varieties. In Swaswa, N, P, and K in the medium reduced by 83.8, 96.2, and 28.7%, while it was 63.3, 61.2, and 23.1% in Kpamyo over the ten-week period. Reduction in K concentration at two weeks was limited to Swaswa. In Kpamyo, adding P at two weeks and N at six weeks would be beneficial. Thus, it is necessary to consider diet requirements from seed production to determine optimum fertigation regimes. This will ensure that plants absorb the available nutrients in sufficient quantities for storage in tubers, which can also be fortified with additional micronutrients if processed into other products after harvest.

7. Outcomes and Impact on Reducing Hunger by Raising Food Security

Food security is commonly defined as the access by all people to sufficient food for active and healthy life (World Bank 1986). Three critical dimensions implicit in this definition are (i) the availability of sufficient quantity and appropriate quality of food supplied through own production or otherwise; (ii) access by all households and individuals to enough and adequate resources to acquire such food; and (iii) the utilization of this food through an adequate diet, water, sanitation, and health care (Timmer 2012). In the developing world, household food security is largely linked to food availability from household production. Gifts and transfers from friends and relatives also play essential roles. Food purchased are also common but limited due to a lack of liquidity.

7.1. Changes Perceived in Household Food Consumption Status of Yam Growing Areas of Nigeria and Ghana

Food deficit was the main periodic shock experienced by most households across the yam-growing areas of Nigeria and Ghana. This type of vulnerability results from qualitative analysis considering the respondents' perception about the number of households affected by food shortages and the frequency of food shortages during a season. To assess the farm household's food consumption during surveys organized by the YIIFSWA project, memory recall on different food shortage scenarios in the past 12 months was employed.

Following the baseline survey, an end-line survey was performed within the same major yam-producing zones in Nigeria and Ghana using the same multistage, random sampling design, drawing on 1400 households based on the same sampling frame (Mignouna et al. 2016). The respondents were asked whether their households had sufficient food during the previous year. Figure 4 shows how households perceived their food security status. This perception was observed to be different between the two rounds of surveys. After the project's first phase, changes in shocks faced by households pursuing their livelihood strategy were noticed by comparing their baseline situation to the end line.

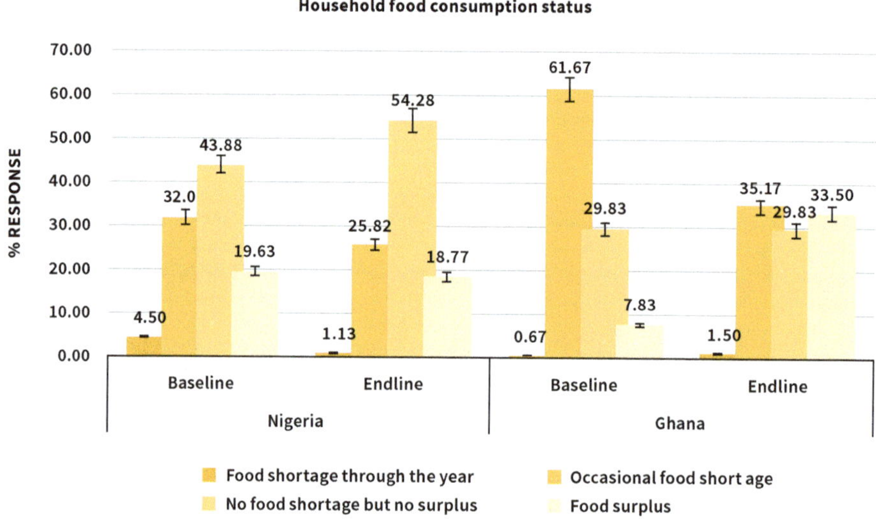

Figure 4. Changes in household food consumption status in a 12-month period for yam growing households in Nigeria and Ghana. Source: (Mignouna et al. 2014a, 2014b, 2017a, 2017b).

In Nigeria, from the baseline to end line, households reported a reduction in food shortages throughout the year from about 5% to 1%. In comparison, reports on occasional food shortages decreased from 32% to about 26% (Figure 4). Households that reported no food shortage, but no surplus increased from about 44% to 54% during the two periods. Households reported that food surplus was almost unchanged during the same periods. In Ghana, the remarkable observations from baseline to the end line are that the proportion of households that reported occasional food shortage decreased by 27% points from about 62% to 35%, while the households that reported having a food surplus increased by 26% from about 8% to 34% (Figure 4). Almost no difference was observed between households that reported no food shortage but no surplus and food shortage throughout the year.

Positive changes in food consumption status were reported, and these probably resulted from increased productivity due to interventions of the YIIFSWA project. They are good indicators of food security improvement in the region, likely due to YIIFSWA's interventions. This translates into a contribution of the project in reducing vulnerability to food insecurity among rural households in the surveyed areas.

7.2. Project Impact on Food Security

Total household expenditure included expenditures on non-food items and consumables. Under food expenditure, all the food items consumed by the household during a year were collected. Food consumption included food that the household purchased, produced, and received from other sources. The total expenditure on food, obtained by aggregating expenditure on all food items, was used to estimate the project's impact on food security. The total expenditure for each food group was calculated by aggregating the expenditure of all food and non-food items falling within a group. The results of the propensity score matching (PSM) presented in Table 2 show that the Average Treatment Effect on the Treated (ATT) of YIIFSWA project on food security (per capita expenditure on food) of beneficiaries in Nigeria and Ghana were about USD 55 and USD 94 for the Radius Matching technique. These results imply that the beneficiaries that profited directly or indirectly from the project are more food secure than the non-beneficiary farmers.

In summary, this study sought to find out the average treatment effect on the treated, which gives the average effect of the project on food security. The results showed a positive impact on food security, implying that the increase in productivity generated by the project interventions led to increased household food security and poverty reduction in the region.

Table 2. Impact of YIIFSWA on food security.

Country	Parameter	YIIFSWA Beneficiaries	YIIFSWA Non-Beneficiaries	Difference	S.E.
		Nearest Neighbor Matching			
Nigeria	Unmatched	348.90	291.10	57.80	17.57
	ATT	350.52	295.54	54.98	19.46
	ATU	290.50	325.00	34.50	.
	ATE			38.16	.
Ghana	Unmatched	715.32	630.36	84.96	179.19
	ATT	718.09	623.95	94.14	175.80
	ATU	635.62	721.47	85.85	.
	ATE			89.41	.

ATT: Average Treatment Effect on the Treated; ATU: Average Treatment on Untreated; ATE: Average Treatment Effect. USD 1 = GHS 1.85 = NGN 157. Source: (Mignouna et al. 2014a, 2014b, 2017a, 2017b).

8. Conclusions

The YIIFSWA initiative has focused on efficient seed yam propagation and production, significantly impacting the yam value chain. The project has developed high-ratio propagation technologies and seed-health management innovations to improve the seed quality and increase productivity by supplying yam farmers with healthy seeds. The new methods emphasize using vines and minitubers of less than 10 g to produce seed tubers and free large quantities of tubers for food. Stakeholders' capacities were enhanced through training on improved seed production methods and equipment to build resilience and sustainability, and increase seed production for income, nutrition, and food security.

9. Recommendations

The gains from the innovations contributed by the YIIFSWA project are visible. However, greater emphasis is required for scaling innovations, especially in certified seed production, for higher impact. Further research efforts are necessary to identify cheaper alternatives in the novel seed yam propagation technologies that will increase the profit margins in seed yam businesses. Studies are also required to identify the most appropriate agronomic practices and control of yam tuber dormancy to allow integration into cropping cycles in the face of production intensification and the changing climate.

Author Contributions: Conceptualization, B.A.; writing—original draft preparation, B.A., R.A.; writing—review and editing, B.A., N.M., R.A., D.M., M.B., P.L.K.; project administration, N.M., R.A.; funding acquisition, IITA: R.A., N.M., P.L.K., M.B., D.M. and B.A. All authors have read and agreed to the published version of the manuscript.

Funding: This research was funded by the Bill and Melinda Gates Foundation, bearing Investment Number OPP1159088. Grant to the International Institute of Tropical Agriculture, Ibadan, Nigeria, November 2016 to 31 December 2021. IITA receives funding for research on yams from the CGIAR Research Program on Roots, Tubers, and Bananas (CRP-RTB) between 2011–2021 and the ongoing CGIAR Seed Equal Initiative, both supported by CGIAR Trust Fund donors.

Conflicts of Interest: The authors declare no conflict of interest. The funders had no role in the design of the study; in the collection, analyses, or interpretation of data; in the writing of the manuscript, or in the decision to publish the results.

References

Adewumi, Adeyinka S., Paterne A. Agre, Paul A. Asare, Michael O. Adu, Kingsley J. Taah, Jean M. Mondo, and Selorm Akaba. 2022. Exploring the *Bush Yam (Dioscorea praehensilis* Benth) as a Source of Agronomic and Quality Trait Genes in White *Guinea Yam (Dioscorea rotundata* Poir) Breeding. *Agronomy* 12: 55. [CrossRef]

Agbaje, Olubunmi G., Lucia O. Ogunsumi, J. A. Oluokun, and T. A. Akinlosutu. 2005. Survey on the Adoption of Yam Minisett Technology in South-Western Nigeria. *Journal of Food, Agriculture and Environment* 3: 222–29.

Aighewi, Beatrice A. 1998. Seed yam (*Dioscorea rotundata Poir.*) Production and Quality in Selected Yam Zones of Nigeria. Ph.D. thesis, Department of Agronomy, Faculty of Agriculture and Forestry, University of Ibadan, Ibadan, Nigeria; p. 252.

Aighewi, Beatrice A., Norbert Maroya, and Robert Asiedu. 2014. *Seed Yam Production from Minisetts: A Training Manual*. Ibadan: International Institute of Tropical Agriculture (IITA), p. 36.

Aighewi, Beatrice A., Norbert Maroya, P. Lava Kumar, Morufat Balogun, Daniel Aihebhoria, Djana Mignouna, and Robert Asiedu. 2021. Seed Yam Production Using High-Quality Minitubers Derived from Plants Established with Vine Cuttings. *Agronomy* 11: 978. [CrossRef]

Aighewi, Beatrice A., Osei Kingsley, Ennin A. Stella, and P. Lava Kumar. 2017. *Practices to Improve the Quality of Seed Yam Produced by Smallholder Farmers. A Seed Yam Production Guide*. Ibadan: International Institute of Tropical Agriculture (IITA), p. 22.

Aighewi, Beatrice A., Robert Asiedu, Norbert Maroya, and Morufat Balogun. 2015. Improved Propagation Methods to Raise the Productivity of Yam (*Dioscorea rotundata* Poir.). *Food Security* 7: 823–34. [CrossRef]

Baah, D. Faustina, Busie Maziya-Dixon, Robert Asiedu, Ibuk Oduro, and William O. Ellis. 2009. Nutritional and Biochemical Composition of *D. alata* (*Dioscorea* spp.) Tubers. *Journal of Food, Agriculture and Environment* 7: 373–78.

Balogun, Morufat O., and Badara Gueye. 2013. Status and Prospects of Biotechnology Applications to Conservation, Propagation and Genetic Improvement of Yam. In *Bulbous Plants: Biotechnology*. Edited by Kishan Gopal Ramawat and Jean-Michel Merillon. New York: CRS Press, pp. 92–112.

Balogun, Morufat, Norbert Maroya, Beatrice Aighewi, Djana Mignouna, Mark Nelson, and Jason Nickerson. 2021. *Manual for Seed yam production in Hydroponics system*. Croydon: International Institute of Tropical Agriculture (IITA) & Context Global Development (CGD), p. 34.

Balogun, Morufat O., Norbert Maroya, Beatrice Aighewi, P. Lava Kumar, Djana Mignouna, Ihuoma Okwuonu, Marian Quain, Adedotun Afolayan, and Emmanuel Chamba. 2020. Capacity strengthening and breeder seed yam production. Paper presented at the 4th YIIFSWA II Annual Review and Planning Meeting, Bolton White Hotel, Abuja, Nigeria, February 18–21; Available online: http://yiifswa.iita.org/wp-content/uploads/2020/04/2019-end-of-year-report-on-pre-basic-seed-production.pdf (accessed on 15 July 2021).

Balogun, Morufat O., Norbert Maroya, Joao Augusto, Adeola Ajayi, Lava Kumar, Beatrice Aighewi, and Robert Asiedu. 2017a. Relative Efficiency of Positive Selection and Tissue Culture for Generating Pathogen-free Planting Materials of Yam (*Dioscorea* spp.). *Czech Journal of Genetics and Plant Breeding* 53: 9–16. [CrossRef]

Balogun, Morufat O., Norbert Maroya, Joao Taiwo, Chukwunalu Ossai, Adeola Ajayi, P. Lava Kumar, Olugboyega Pelemo, Beatrice Aighewi, and Robert Asiedu. 2017b. *Clean Breeder Seed Yam Tuber Production Using Temporary Immersion Bioreactors*. Ibadan: International Institute of Tropical Agriculture (IITA), p. 66. ISBN 978-978-8444-89-3.

Craufurd, Peter O., Rodney J. Summerfield, Robert Asiedu, and Vara Prasad. 2001. Dormancy in Yams. *Journal of Experimental Agriculture* 37: 75–109. [CrossRef]

Dave, Buenavista P., Nikko Manuel A. Dinopol, Eefke Mollee, and Morag McDonald. 2020. From Poison to Food: On the Molecular Identity and Indigenous Peoples' Utilisation of Poisonous "Lab-o" (Wild Yam, Dioscoreaceae) in Bukidnon, Philippines. *Cogent Food & Agriculture* 7: 1870306. [CrossRef]

Elbehri, Aziz. 2013. *West Africa's Food Potential: Policies and Market Incentives for Smallholder-inclusive Food Value Chains*. Rome: The Food and Agriculture Organization of the United Nations and the International Fund for Agriculture Development.

FAOSTAT. 2021. Food and Agriculture Organization of the United Nations. On-Line and Multilingual Database. Available online: http://faostat.fao.org/ (accessed on 30 November 2021).

Frossard, Emmanuel, Beatrice A. Aighewi, Sévérin Aké, Dominique Barjolle, Philipp Baumann, Thomas Bernet, Daouda Dao, Lucien N. Diby, Anne Floquet, Valérie K. Hgaza, and et al. 2017. The Challenge of Improving Soil Fertility in Yam Cropping Systems of West Africa. *Frontiers in Plant Science* 8: 1953. [CrossRef]

Hollinger, Frank, and John M. Staatz. 2015. *Agricultural Growth in West Africa Market and Policy Drivers*. Rome: African Development Bank and the Food and Agriculture Organization of the United Nations.

Joob, Beuy, and Viroj Wiwanitkit. 2014. Asiatic Bitter Yam Intoxication. *Asian Pacific Journal of Tropical Biomedicine* 4: S42. [CrossRef]

Kyung-Sik, Kang, and S. Taek Heo. 2015. A Case of Life-threatening Acute Kidney Injury with Toxic Encephalopathy caused by *Dioscorea quinqueloba*. *Yonsei Medical Journal* 56: 304–6. [CrossRef]

Lev, Larry S., and Ann L. Shriver. 1998. A Trend Analysis of Yam Production, Area, Yield, and Trade (1961–1996). In *Actes du Seminaire International CIRAD-INRA-ORSTOM-CORAF CIRAD*. Edited by Julien Berthaud, Nicolas Bricas and Jean-Leu Marchand. Montpellier: L'igname, Plante Seculaire et Culture d'avenir, pp. 13–20. Available online: https://horizon.documentation.ird.fr/exl-doc/pleins_textes/divers17-08/010014891.pdf (accessed on 30 November 2021).

Manyong, M. Victor. 2000. Farmers' Perceptions of the Resource Management Constraints in Yam-based Systems. Project 13: Improvement of Yam-Based Systems. In *Annual Report 1999*. Ibadan: IITA, pp. 3–4.

Maroya, Norbert, Morufat Balogun, and Robert Asiedu. 2014a. Seed yam production in an aeroponics system: A novel technology. In *YIIFSWA Working Paper Series No 2 (Revised). Yam Improvement for Income and Food Security in West Africa*. Ibadan: International Institute of Tropical Agriculture (IITA), p. 20.

Maroya, Norbert, Morfa Balogun, Beatrice Aighewi, Djana B. Mignouna, P. Lava Kumar, and Robert Asiedu. 2022. Transforming Yam Seed Systems in West Africa. In *Root, Tuber and Banana Food System Innovations*. Edited by Graham Thiele, Michael Friedmann, Hugo Campos, Vivian Polar and Jeffery W. Bentley. Cham: Springer. [CrossRef]

Maroya, Norbert, Morufat Balogun, Beatrice Aighewi, Jimoh Lasisi, and Robert Asiedu. 2017. *Manual for Clean Foundation Seed Yam Production Using Aeroponics System*. Ibadan: International Institute of Tropical Agriculture (IITA), p. 68. ISBN 978-978-8444-88-6.

Maroya, Norbert, Morufat Balogun, Robert Asiedu, Beatrice Aighewi, P. Lava Kumar, and Joao Augusto. 2014b. Yam Propagation Using Aeroponics Technology. *Annual Research and Review in Biology* 4: 3894–903. [CrossRef]

Maroya, Norbert, Robert Asiedu, P. Lava Kumar, Djana Mignouna, Antonio Lopez-Montes, Ulrich Kleih, David Phillips, Fadel Ndiame, John Ikeorgu, and Emmanuel Otoo. 2014c. Yam Improvement for Income and Food Security in West Africa: Effectiveness of a Multi-Disciplinary and Multi-Institutional Team-Work. *Journal of Root Crops* 40: 85–92.

Mignouna, Djana B., Adebayo Akinola, Tahirou Abdoulaye, Arega Alene, Norbert Maroya, Beatrice Aighewi, Thomas Wobill, and Robert Asiedu. 2016. *YIIFSWA Working Paper Series No 7. Impact Evaluation Protocols for Agricultural Projects: The Case of Yam Improvement for Income and Food Security in West Africa, Impact-Evaluation Guidelines*. Ibadan: International Institute of Tropical Agriculture (IITA), p. 70.

Mignouna, Djana B., Akinola Adebayo, Tahirou Abdoulaye, Arega Alene, Victor Manyong, Norbert Maroya, Beatrice Aighewi, Lava P. Kumar, Balogun Morufat, Antonio Lopez-Montes, and et al. 2020. Potential Returns to Yam Research Investment in Sub-Saharan Africa and Beyond. *Outlook on Agriculture* 49: 215–24. [CrossRef]

Mignouna, Djana B., Maroya Norbert, Aighewi Beatrice, Kumar Lava, Akinribido Bolanle, Ikeorgu John, Abdoulaye Tahirou, Akinola Adebayo, Alene Arega, Manyong Victor, and et al. 2017a. *Impact Assessment Report of YIIFSWA Project. Raising Household Income, Improving Food Security and Reducing Poverty in Nigeria*. Ibadan: International Institute of Tropical Agriculture (IITA), p. 48. ISBN 978-978-8444-94-7.

Mignouna, Djana B., Maroya Norbert, Aighewi Beatrice, Kumar Lava, Abdoulaye Tahirou, Akinola Adebayo, Otabil Prince, Braimah Haruna, Alene Arega, Manyong Victor, and et al. 2017b. *Impact Assessment Report of YIIFSWA Project. Raising Household Income, Improving Food Security and Reducing Poverty in Ghana*. Ibadan: International Institute of Tropical Agriculture (IITA), p. 48. ISBN 978-978-8444-95-4.

Mignouna, Djana B., Tahirou Abdoulaye, and Adebayo Akinola. 2015. Factors Influencing the Use of Selected Inputs in Yam Production in Nigeria and Ghana. *Journal of Agriculture and Rural Development in the Tropics and Subtropics* 116: 131–42.

Mignouna, Djana B., Tahirou Abdoulaye, Arega Alene, Robert Asiedu, and Victor M. Manyong. 2014a. *Characterization of Yam-growing Households in the Project Areas of Ghana*. Ibadan: International Institute of Tropical Agriculture (IITA), p. 95. ISBN 978-978-8444-45-9.

Mignouna, Djana B., Tahirou Abdoulaye, Arega Alene, Robert Asiedu, and Victor M. Manyong. 2014b. *Characterization of Yam-Growing Households in the Project Areas of Nigeria*. Ibadan: International Institute of Tropical Agriculture (IITA), p. 92. ISBN 978-978-8444-46-6.

Mondo, Jean M., Paterne A. Agre, Alex Edemodu, Patrick Adebola, Robert Asiedu, Malachy O. Akoroda, and Asrat Asfaw. 2020. Floral Biology and Pollination Efficiency in Yam (*Dioscorea* spp.). *Agriculture* 10: 560. [CrossRef]

Ngo, Ngwe M.F.S., N. Denis Omokolo, and Simon Joly. 2015. Evolution and Phylogenetic Diversity of Yam Species (*Dioscorea* spp.): Implication for Conservation and Agricultural Practices. *PLoS ONE* 10: e0145364. [CrossRef]

Nweke, Felix I. 2016. *Yam in West Africa: Food, Money and More*. East Lansing: Michigan State University Press.

Obidiegwu, Jude E., and Emmanuel M. Akpabio. 2017. The Geography of Yam Cultivation in Southern Nigeria: Exploring its Social Meanings and Cultural Functions. *Journal of Ethnic Foods* 4: 28–35. [CrossRef]

Obidiegwu, Jude E., Jessica B. Lyons, and Cynthia A. Chilaka. 2020. The *Dioscorea* Genus (Yam)—An Appraisal of Nutritional and Therapeutic Potentials. *Foods* 9: 1304. [CrossRef]

Orkwor, Gabriel C., and Charles L. A. Asadu. 1998. Agronomy. In *Food Yams. Advances In Research*. IITA and NRCRI, Nigeria. Edited by G. C. Orkwor, R. Asiedu and I. J. Ekanayake. Ibadan: INTEC Printers, pp. 105–41.

Price, Elliott J., Paul Wilkin, Viswambharan Sarasan, and Paul D. Fraser. 2016. Metabolite Profiling of Dioscorea (Yam) Species Reveals Underutilised Biodiversity and Renewable Sources for High-value Compounds. *Scientific Reports* 6: 29136. [CrossRef]

Shajeela, P. S., Veerabahu R. Mohan, Jesudas L. Louis, and Soris P. Tresina. 2011. Nutritional and Antinutritional Evaluation of Wild Yam (*Dioscorea* spp.). *Tropical and Subtropical Agroecosystems* 14: 723–30.

Srivastava, Amit K., Thomas Gaiser, Heiko Paeth, and Frank Ewert. 2012. The Impact of Climate Change on Yam (*Dioscorea alata*) Yield in the Savanna Zone of West Africa. *Agriculture, Ecosystems and Environment* 153: 57–64. [CrossRef]

Timmer, Peter. 2012. Behavioral Dimensions of Food Security. *Proceedings of the National Academy of Sciences* 109: 12315–20. [CrossRef] [PubMed]

Tohda, Chihiro, Ximeng Yang, Mie Matsui, Yuna Inada, Emika Kadomoto, Shotaro Nakada, Hidetoshi Watari, and Naotoshi Shibahara. 2017. Diosgenin-rich Yam Extract Enhances Cognitive Function: A Placebo-controlled, Randomized, Double-blind, Crossover Study of Healthy Adults. *Nutrients* 9: 1160. [CrossRef] [PubMed]

van Etten, Jacob, Isabel L. Noriega, Carlo Fadda, and Evert Thomas. 2017. The Contribution of Seed Systems to Crop and Tree Diversity in Sustainable Food Systems. Available online: https://cgspace.cgiar.org/handle/10568/89755 (accessed on 15 July 2021).

World Bank. 1986. *Poverty and Hunger: Issues and Options for Food Security in Developing Countries*. Washington, DC: World Bank.

YIIFSWA. 2017. *Yam Improvement for Income and Food Security in West Africa Project Achievements and Recommendations Report*. Submitted on 28 February 2017 to BMGF.

Yoon, Jae C., Jae B. Lee, Tae O. Jeong, Si O. Jo, and Young H. Jin. 2019. Symptomatic Hypocalcemia Associated with Dioscorea tokoro Toxicity. *Journal of the Korean Society of Clinical Toxicology* 17: 42–45. [CrossRef]

Coconut-Based Livestock Farming: A Sustainable Approach to Enhancing Food Security in Sri Lanka

Tharindu D. Nuwarapaksha, Shashi S. Udumann, Nuwandhya S. Dissanayaka and Anjana J. Atapattu

1. Introduction

Coconut (*Cocos nucifera* L.) has emerged as a versatile plantation crop, carrying crucial nutritional value and boundless functionalities that open doors to various industrial undertakings (Nair 2021). With the diverse applications of its different parts, the coconut palm has become an inseparable aspect of the social, economic, and cultural lives of nearly 80 million people across 92 countries (Nair 2021; Beveridge et al. 2022). The history of coconut cultivation in Sri Lanka can be traced back to 330 B.C. The crop is grown across twenty-five administrative districts on the island. In many aspects, coconut cultivation bears similarities to the extension and historical significance of paddy cultivation. Regarding land coverage, coconut ranks second, following the country's staple food crop, paddy, accounting for approximately 20% of the total agricultural extent on the island. The coconut industry also contributes to 0.7% of the Gross Domestic Product (GDP), with the total extent of coconut cultivation in the country being 505,000 hectares (CBSL 2021). Annual coconut production fluctuates yearly due to various factors, such as weather conditions, disease outbreaks, and market demand (Utomo et al. 2016). Approximately 58% of coconut cultivation is concentrated in three districts: Gampaha, Kurunegala, and Puttalam, which are collectively known as the "Coconut Triangle" (Esham and Wijeratne 2021). Regarding the share of the total coconut extent compared to the total agricultural extent by district, Puttalam district stands at the top, with approximately 69.8% of its land devoted to coconut cultivation. Gampaha district follows closely at 64.0%, and Kurunegala district is at 60.6% (DCS 2021). The lowest share is reported in Nuwara Eliya district, where only 0.6% of the land is allocated to coconut cultivation due to the dominance of tea as the primary crop.

Moreover, animal husbandry plays a crucial role in the Sri Lankan agricultural sector, with a focus on raising various types of animals. Similar to coconut farming, the livestock sector makes a comparable contribution of approximately 0.75% to the total GDP (CBSL 2021). The main livestock raised in the country include cattle, water buffalo, goats, sheep, pigs, and poultry (Perera and Jayasuriya 2008). Cattle and water buffalo are primarily bred for milk production, making a significant contribution to the dairy sector. On the other hand, goats and sheep are versatile animals used for milk and meat production (Rout and Behera 2021). In Sri Lanka, traditional animal husbandry practices typically revolve around small-scale, mixed crop–livestock farming systems. However, coconut plays a significant role as a plantation crop in Sri Lanka, and due to its specific morphological characteristics, only 25% of this land is utilized for monoculture coconut cultivation (Samarajeewa et al. 2003; Nuwarapaksha et al. 2022).

The practice of incorporating livestock into coconut farming to achieve maximum land use efficiency is becoming popular among coconut farmers. This is primarily because of its potential to improve resource sharing, especially in areas with limited land for intercropping on coconut farms (Nuwarapaksha et al. 2023a). This presents opportunities for the integration of livestock or intercropping practices on coconut land. According to Ibrahim and Jayatileka (2000), approximately 22% of natural grasslands exist on coconut land; therefore, it has been identified as a prime location for the development of livestock activities. The abundance of coconut plantations in Sri Lanka presents an excellent opportunity for implementing sustainable agricultural practices and advancing livestock development (Nuwarapaksha et al. 2022). Leveraging the inherent strengths of Sri Lanka, the cultivation of forage crops in coconut plantations stands as the key substantial agricultural endeavor (Table 1). Recognizing the capabilities of the nation, strategically incorporating livestock or intercropping practices within these coconut-rich areas presents a noteworthy opportunity to advance agricultural sustainability across the country.

The intersection of coconut farming and animal husbandry assumes a key role in ensuring food security, as noted by Ansar and Fathurrahman (2018). The deep involvement of the rural population in coconut cultivation and other agroforestry pursuits establishes a robust link to the production of food, wood, energy resources, and animal feed. Additionally, forests contribute significantly to vital ecological functions, encompassing carbon storage, the preservation of wildlife habitats, and the protection of environmental resources. Consequently, these multifaceted contributions offer a tangible pathway toward achieving the SDGs (Ruba and Talucder 2023). In the pursuit of achieving the SDGs, they have been categorized into five distinct groups, each dedicated to addressing specific thematic areas. These categories include Category 1 (encompassing SDGs 1–5, addressing poverty), Category 2 (covering SDGs 6–9, related to development infrastructure), Category

3 (addressing SDGs 10–12, focusing on sustainable production and consumption), Category 4 (encompassing SDGs 13–15, related to green infrastructure), and Category 5 (incorporating SDGs 16–17, concerning green institutions) (Sharma et al. 2022). Within the context of coconut-based livestock farming systems, it has been observed that Category 1 and SDG 2 have been particularly well realized (McElwee et al. 2020). Hence, the primary objective of this chapter is to investigate the concept of coconut-based livestock farming in the Sri Lankan context, aiming to contribute to the realization of the SDGs, particularly in the realm of ensuring food security.

Table 1. Forage sources, land extent, and expected production in Sri Lanka.

Source	Approx. Land Extent (000′ Hectare)	Expected Dry Matter Production (MT′)			
		Favorable (Rainy Season)		Unfavorable (Drought Season)	
		MT′ Per ha and Single Defoliation	Total * MT	MT′ Per ha and Single Defoliation	Total ** MT
Coconut plantations (30% of total coconut lands)	145	1.25	543,750	0.65	188,500
Permanent pasture under large farms—well managed pasture	5.7	2.5	42,750	2	22,800
Permanent pasture under large farms—poorly managed pasture	13.27	1.25	49,763	0.65	17,251
Permanent fodder under large farms—well managed	1.87	7.5	42,075	6.5	24,310
Cultivated fodder under Small and medium holders	18.5	7.5	416,250	4.5	199,800
Marginal tea lands	30.23	0.75	68,017	0.37	22,370
Fellow paddy fields	150	1.5	675,000	1	300,000
Road siders/railway embankments	5	1.5	22,500	0.75	7500
All types of natural grasslands/wastelands (50% total land extent)	530	0.75	1,192,500	0.37	392,200
Total	899.57		3,052,605		1,174,731

* Minimum 3 defoliations during the season for the total land extent; ** minimum 2 defoliations during the season for total land extent. Source: Authors' compilation based on data from Weerasinghe (2019).

2. Coconut-Based Livestock Farming Systems

The coconut-based livestock farming system involves integrating livestock and crops on the same land to establish a mutually beneficial relationship between them (Devendra 2007). This approach can take various forms, including agroforestry, silvopastoral systems, mixed crop–livestock systems, and integrated crop–livestock systems, all characterized by the interdependence of their components (Altieri et al. 2015). By combining crops and animals, farmers can achieve several advantages, such as diversifying production, lessening reliance on external inputs, enhancing resilience to climate variations and market fluctuations, and increasing overall income. Moreover, this integrated approach fosters improved soil health, biodiversity, reduced greenhouse gas emissions, and more effective waste management (Lehmann et al. 2020).

2.1. Coconut-Based Agroforestry Systems

Coconut-based agroforestry refers to a land management practice that combines the cultivation of coconut trees with the integration of other tree species, agricultural crops, and livestock in the same farming system (Dissanayaka et al. 2023). This agroforestry approach aims to optimize the utilization of land, enhance overall productivity, and promote environmental sustainability by harnessing the benefits of diverse components (Atapattu et al. 2017). In addition to coconut trees, other crop species are strategically planted within the coconut plantation (Figure 1). These companion crops can include fruit crops, vegetable crops, export agriculture crops, forages, timber crops, or legume crops that contribute to soil fertility and ecosystem health (Nuwarapaksha et al. 2022). The combination of different tree species provides a range of ecological and economic benefits, such as improving biodiversity, enhancing carbon sequestration, and generating additional income from various tree products (Jose 2009). The overarching goal is to maximize the social, economic, and environmental benefits for land users at various levels. Agroforestry plays a significant role in climate-smart agriculture, serving both adaptation and mitigation purposes (Dawid Mume and Workalemahu 2021). Trees and agroforestry systems offer a diverse array of goods and services that can serve as substitutes for one another and, under suitable conditions, can be produced in a synergistic manner. Livestock-keeping serves as a means to diversify the production options of rural communities and is frequently suited to marginal environments. This adaptability can contribute to enhancing climate resilience within these communities (Behera and France 2016). The interaction between agroforestry and livestock-keeping is vital for the livelihoods of rural communities, and the combination of these two practices plays a crucial role in supporting and sustaining the well-being of rural communities.

Figure 1. Components of coconut-based livestock farming systems: (**a**) coconut + gliricidia + pepper mixed cropping; (**b**) fodder sorghum grass cultivation under coconut plantation; (**c**) goat naturally grazing on coconut land; (**d**) biogas production under coconut plantation. Source: Figures by authors.

2.2. *Coconut-Based Silvopastoral Systems*

Silvopastoral farming is a popular practice involving cultivating a harmonious combination of grasses, legumes, and trees while raising livestock on the same parcel of land (Kreitzman et al. 2022). Its aim to optimize land productivity, conserve soil and water, and yield a variety of resources such as forage, fuelwood, and timber on a sustainable basis. Planting trees and understory components is essential to create a thriving silvopastoral system (Gautam et al. 2003). The main difference between agroforestry and silvopastoral systems is their emphasis and scope. Agroforestry encompasses a broader range of interactions between trees and various agricultural or livestock activities, while silvopastoral systems specifically focus on integrating fodder grasses with livestock grazing (Soler et al. 2018). Mainly forages and shrubs are planted to feed the animals, and the livestock manure is used as crop fertilizer (Figure 1). The integration of trees, grasses, and legumes (gliricidia) in silvopasture contributes to soil and moisture conservation (Raveendra et al. 2021). The specific systems and practices are identified based on the trees' or shrubs' role in the setup (Leakey 2017). The successful establishment of silvopasture involves the careful selection of suitable tree species, shrubs, grasses, and legumes suited to different regions. Additionally, measures to protect the area from animals and appropriate soil and water conservation methods are crucial. However, a well-designed and properly managed silvopastoral system aligns closely with the goals of SDG 2 by addressing hunger, promoting sustainable agriculture, and improving food security and nutrition. Its multifunctional benefits contribute to a more resilient and holistic approach to meeting the global challenges of food production and access.

2.3. Coconut-Based Mixed Crop–Livestock Systems

The mixed crop–livestock system focuses on integrating crops and livestock (Thornton and Herrero 2014). This system aims to achieve mutual benefits between crops and livestock, each supporting and complementing the other regarding resources, productivity, and sustainability. In mixed crop–livestock systems, farmers strategically plan the integration of crops and livestock based on their complementary interactions (Figure 1). For instance, crop residues, such as leftover plant materials after harvest, can serve as valuable feed for livestock, reducing the need for additional fodder. In return, the manure produced by the livestock serves as a natural fertilizer for the crops, enhancing soil fertility and nutrient cycling. The diversity introduced by mixed crop–livestock systems contributes to improved resilience against climatic variability and potential economic risks (Thornton and Herrero 2014). Farmers are less dependent on a single income source, as they can derive revenue from both crop sales and livestock products, such as meat, milk, and wool. Crop–livestock integration also helps in managing pests and diseases. For example, some livestock species can graze on weeds that would otherwise compete with crops for resources or act as hosts for pests (Hilimire 2011). Overall, mixed crop–livestock systems offer a promising approach to sustainable and resilient agriculture, providing a balance between food production, natural resource conservation, and economic stability for farming communities (Sekaran et al. 2021). By harnessing the benefits of crops and livestock, these integrated systems contribute to promoting food security and sustainable development in various regions across the world.

2.4. Coconut-Based Integrated Crop–Livestock Systems

The core component of these systems is coconut cultivation, and they are typically combined with other crops such as intercropping of food crops or forages beneath the coconut canopy. Livestock may also be reared on the same land.

This practice is known as intercropping or mixed cropping and allows for the simultaneous cultivation of food crops, vegetables, or forages, which complement coconut farming (Sekaran et al. 2021). Coconut-based integrated crop–livestock systems are innovative and sustainable agricultural practices that revolve around the cultivation of coconut trees alongside the integration of diverse crops and livestock (Wulandari 2021), while biogas production is a crucial and necessary component in an integrated crop–livestock system (Figure 1). It plays a pivotal role in converting organic waste, such as livestock manure and agricultural residues, into valuable biogas and nutrient-rich digestate. This biogas serves as a renewable energy source, while the digestate can be used as a natural fertilizer for crops, completing the cycle of sustainability (Surendra et al. 2014). By utilizing the anaerobic digestion process, methane-rich biogas can be extracted from organic waste materials, which can then be harnessed for cooking, heating, and electricity generation in daily life. The diverse array of crops adds to the overall farm productivity and provides additional income sources. Furthermore, coconut-based integrated crop–livestock systems often include livestock rearing as an integral component (Seresinhe and Sujani 2016). The well-managed integration of coconut-based cultivation with livestock rearing epitomizes harmonious alignment with the objectives of SDG 2.

3. Fodder Production and Conservation on Coconut Land

Fodder crops play a significant role in enhancing livelihoods through diverse means. The primary ways in which fodder crops contribute to improved food security, increased incomes, and overall better livelihoods have been demonstrated (Franzel et al. 2014). Apart from naturally occurring and naturalized varieties like guinea grass, there are a few cultivated pasture and fodder varieties in Sri Lanka. The most commonly established variety is the hybrid Napier (CO3) (Weerasinghe 2019). CO3 is specifically grown for livestock use and receives government support through the Department of Animal Production and Health and its respective provincial authorities. It is primarily cultivated for farmers' own use, although there are a few instances where it is grown for sale outside the farm. In addition to CO3, CO4, CO5, and Red Napier, are also gaining popularity among farmers as improved fodder varieties (Figure 2). Additionally, farmers cultivate fodder sorghum and fodder maize to feed the livestock. Additionally, gliricidia and ipil-ipil are also grown, primarily for use as boundary fencing and as feed for ruminant animals (Weerasinghe 2019; Nuwarapaksha et al. 2023b).

Figure 2. Fodder grasses: (**a**) CO3; (**b**) CO4; (**c**) CO5; and (**d**) Red Napier cultivation under coconut plantation. Source: Figures by authors.

The primary method of conserving fodder in Sri Lanka is through the use of silage (Figure 3). Previous attempts to produce hay using readily available grasses have been made through various programs, but these efforts have been unsuccessful thus far. The lack of commercial or large-scale pasture production in the country, unfavorable weather conditions (such as heavy rain during the two monsoon seasons affecting haymaking, and a scarcity of good-quality pasture during the dry season), and high operational costs are potential reasons for this lack of success. In contrast, silage-making has become viable for commercial dairy farms and small-scale on-farm use. There are three main methods of silage production: small-scale silage production using plastic barrels or vacuum-packed polythene bags, bunker silage production at large-scale commercial dairy farms, and commercial silage production, mainly in bales (Weerasinghe 2019). The crops used for this purpose are maize and/or sorghum; typically, the silage is produced for on-farm use.

Figure 3. Small-scale barrel silage production from fodder maize grasses. Source: Figures by authors.

4. Contributions of Coconut-Based Livestock Farming to SDGs

Coconut-based livestock farming can contribute significantly to achieving several SDGs, including SDG 2. Both coconut cultivation and animal husbandry practices can improve the productivity of animals and coconut land' and the quality of the products produced from them (Devendra 2007). They can also play a role in sustainable food production and preserving genetic diversity, especially in relation to eradicating hunger, promoting sustainable agriculture, and ensuring the well-being of communities. The following SDGs can be achieved mainly through coconut-based livestock farming systems (Figure 4). By using coconut by-products as livestock feed and integrating coconut farming with animal husbandry, we can enhance the availability of nutritious food for both humans and animals. Increased livestock productivity can provide a sustainable source of foods and beverages, contributing to food security and reducing hunger. Coconut-based livestock farming can enhance food security and nutrition by providing a diversified source of food products, such as milk, meat, and eggs, to meet the dietary needs of the population (SDG 2) (Paramesh et al. 2022).

Integrating livestock activities within coconut plantations can create employment opportunities, particularly in rural areas, leading to economic growth and poverty reduction (SDG 8) (Devendra and Chantalakhana 2002). By adopting sustainable practices in coconut-based livestock farming, such as efficient resource use, reduced waste, and environmentally friendly methods, this approach can contribute to responsible consumption and production patterns (SDG 12) (Behera and France 2016). Sustainable coconut-based livestock farming practices can help mitigate climate change by promoting carbon sequestration through coconut trees and implementing climate-resilient farming techniques (SDG 13) (Nair et al.

2018). Integrating livestock within coconut plantations can support biodiversity conservation and land restoration, contributing to the protection of terrestrial ecosystems (SDG 15) (Zoysa and Inoue 2014). Collaboration between the government, the private sector, local communities, and international organizations is essential to realizing the potential of coconut-based livestock farming in achieving the SDGs (SDG 17) (Achmad et al. 2022). By aligning coconut-based livestock farming with these SDGs, Sri Lanka can harness the full potential of its coconut resources while promoting sustainable and inclusive agricultural practices that benefit both people and the planet.

Figure 4. Visual concept: contributions of coconut-based livestock farming to SDGs. Source: Figure by authors.

5. Benefits of Coconut-Based Livestock Farming Systems

Coconut-based farming systems, integrating coconut plantations with livestock and other vegetation, offer a myriad of advantages. They improve livestock nutrition by providing diverse fodder options within the coconut plantation, enhancing animal health (Devendra and Leng 2011). These systems optimize land use by combining grazing and coconut cultivation on the same land, leading to efficient resource utilization and reducing the need for additional grazing areas. Moreover, livestock grazing within the plantation aids in weed and pest control, minimizing the use of herbicides and maintaining pest balance (Nicholls and Altieri 2013). Furthermore, the integration of nitrogen-fixing trees, including gliricidia and shrubs, enhances soil fertility, benefiting coconut trees and other crops (Raveendra et al. 2021). Silvopastoral systems also contribute to carbon sequestration through the excellent CO_2 absorption capabilities of coconut trees and diversified vegetation (Ramachandran Nair et al. 2010). According to Raveendra et al. 2017, carbon stocks in a gliricidia-based mixed cropping system was significantly different compared to that in a mono-crop system. They also promote biodiversity, ecosystem resilience, and improved pest and disease management. Farmers gain multiple income streams from coconut products, livestock, and other agricultural produce, enhancing their financial and energy security. Integrated systems foster climate adaptation by providing a resilient and adaptable farming approach, helping farmers cope with climate-related challenges. They offer a sustainable and integrated agriculture model, delivering numerous environmental, economic, and social benefits to farmers and the surrounding ecosystem (Hernández-Morcillo et al. 2018). In summary, coconut-based livestock farming can contribute to achieving SDG 2, which aims to end hunger, achieve food security, improve nutrition, and promote sustainable agriculture.

6. Challenges for Coconut-Based Livestock Farming Systems

Coconut-based livestock farming systems offer numerous benefits, but also present challenges that require careful attention from farmers and land managers to ensure success and sustainability. Integrating livestock and vegetation within coconut plantations can create competition for vital resources like water, sunlight, and nutrients, necessitating a delicate balance to maintain productivity (Kumar and Kunhamu 2022). Selecting compatible livestock breeds becomes crucial to avoid potential damage to coconut trees and maintain a thriving environment. Managing pests and diseases in this complex system requires vigilant monitoring and timely intervention. Successful implementation demands a deep understanding of agroforestry practices, livestock management, and coconut cultivation, emphasizing the need for access to proper training and knowledge (Mizik 2021). Long-term planning and investments in planting fodder trees, acquiring suitable livestock, and diligent management are essential for achieving sustainable benefits. Market access and fair prices for coconut and livestock products play a pivotal role in ensuring economic viability (Kaplinsky and Morris 2018). Overcoming land tenure issues and obtaining policy support are vital for encouraging farmers to adopt these integrated systems. Climate variability and environmental risks necessitate climate-resilient strategies. Additionally, labor-intensive management is required compared to conventional monoculture plantations (Feintrenie et al. 2015). Despite these challenges, with careful planning, training, and management, coconut-based farming systems can flourish, offering a sustainable and resilient agricultural approach that benefits farmers and the environment alike.

7. Conclusions

Coconut-based livestock farming represents a promising pathway to achieve several SDGs outlined by the United Nations, mainly SDG 2, as well as SDG 8, SDG 12, SDG 13, SDG 15, and SDG 17. By integrating coconut plantations with livestock and other vegetation, this innovative agroforestry approach offers a range of benefits, promoting environmental sustainability, economic growth, and social well-being. Coconut-based livestock farming offers a promising and sustainable approach to agricultural production, combining the benefits of coconut cultivation with livestock rearing. This integrated farming system provides numerous advantages for farmers, the environment, and local communities. By strategically managing coconut plantations alongside livestock, farmers can diversify their income sources and improve overall farm productivity. The system promotes biodiversity by creating diverse habitats for various plant and animal species. By diversifying production and income sources, farmers are better equipped to withstand challenges and ensure the well-being of their communities. Overall, coconut-based livestock farming represents a holistic and ecologically sound approach to agriculture in Sri Lanka. By capitalizing on the strengths of both coconut cultivation and livestock rearing, this integrated farming system offers a pathway to sustainable and prosperous farming practices. As the world faces increasing demands for food production and environmental stewardship, coconut-based livestock farming stands as a viable model to support the livelihoods of farmers while conserving natural resources for future generations. Further research and development programs are very important in determining the most suitable crop species and suitable livestock species for each system.

Author Contributions: Conceptualization, T.D.N. and A.J.A.; Methodology, S.S.U.; Validation, A.J.A. and N.S.D.; Formal Analysis, T.D.N.; Investigation, S.S.U.; Writing—Original Draft Preparation, T.D.N. and S.S.U.; Writing—Review and Editing, A.J.A. and N.S.D.; Supervision, A.J.A.; Visualization, T.D.N. and S.S.U.; Project Administration, A.J.A. All authors contributed to the article and approved the submitted version.

Funding: This research received no external funding.

Acknowledgments: We would like to extend our gratitude to the technical team at the Agronomy Division of the Coconut Research Institute for their valuable contributions. We also want to express our thanks to the editor and two anonymous reviewers for providing constructive feedback and valuable comments.

Conflicts of Interest: The authors declare no conflict of interest.

References

Achmad, Budiman, Sanudin, Mohamad Siarudin, Ary Widiyanto, Dian Diniyati, Aris Sudomo, Aditya Hani, Eva Fauziyah, Endah Suhaendah, Tri Suslistyati Widyaningsih, and et al. 2022. Traditional Subsistence Farming of Smallholder Agroforestry Systems in Indonesia: A Review. *Sustainability* 14: 8631. [CrossRef]

Altieri, Miguel A., Clara I. Nicholls, Alejandro Henao, and Marcos A. Lana. 2015. Agroecology and the Design of Climate Change-Resilient Farming Systems. *Agronomy for Sustainable Development* 35: 869–90. [CrossRef]

Ansar, Muhammad, and Fathurrahman. 2018. Sustainable integrated farming system: A solution for national food security and sovereignty. *IOP Conference Series: Earth and Environmental Science* 157: 012061. [CrossRef]

Atapattu, Anjana J., Sumith H. S. Senarathne, Thilina Raveendra, Chaminda Egodawatte, and Sylvanus. Mensah. 2017. Effect of Short-Term Agroforestry Systems on Soil Quality in Marginal Coconut Lands in Sri Lanka. *Agricultural Research Journal* 54: 324. [CrossRef]

Behera, Umakanta, and James France. 2016. Integrated Farming Systems and the Livelihood Security of Small and Marginal Farmers in India and Other Developing Countries. *Advances in Agronomy* 138: 235–82. [CrossRef]

Beveridge, Fernanda Caro, Sundaravelpandian Kalaipandian, Chongxi Yang, and Steve W. Adkins. 2022. Fruit Biology of Coconut (*Cocos nucifera* L.). *Plants* 11: 3293. [CrossRef]

CBSL. 2021. Central Bank Annual Report Data. Ministry of finance. Colombo. Available online: https://www.cbsl.gov.lk/sites/default/files/cbslweb_documents/publications/annual_report/2021/en/Full_Text_Volume_I.pdf (accessed on 13 July 2023).

Dawid Mume, Ibsa, and Sisay Workalemahu. 2021. Review on Windbreaks Agroforestry as a Climate Smart Agriculture Practices. *American Journal of Agriculture and Forestry* 9: 342. [CrossRef]

DCS. 2021. Department of Census and Statistics Sri Lanka Annual Report Data. Ministry of Finance. Colombo. Available online: http://www.statistics.gov.lk/Agriculture/StaticalInformation/rub4 (accessed on 8 August 2023).

Devendra, Canagasaby. 2007. Perspectives on Animal Production Systems in Asia. *Livestock Science* 106: 1–18. [CrossRef]

Devendra, Canagasaby, and Charan Chantalakhana. 2002. Animals, Poor People and Food Insecurity: Opportunities for Improved Livelihoods through Efficient Natural Resource Management. *Outlook on Agriculture* 31: 161–75. [CrossRef]

Devendra, Canagasaby, and Ronald Alfred Leng. 2011. Feed Resources for Animals in Asia: Issues, Strategies for Use, Intensification and Integration for Increased Productivity. *Asian-Australasian Journal of Animal Sciences* 24: 303–21. [CrossRef]

Dissanayaka, Nuwandhya S., Lakmini Dissanayake, Shashi S. Udumann, Tharindu D. Nuwarapaksha, and Anjana J. Atapattu. 2023. Agroforestry a Key Tool in the Climate-Smart Agriculture Context: A Review on Coconut Cultivation in Sri Lanka. *Frontiers in Agronomy* 5: 1162750. [CrossRef]

Esham, Mohamed, and A. W. Wijeratne. 2021. Risk and Vulnerability Assessment for Resilience Building in the Colombo City Region Food System (CRFS). Sustainability and Climate Change Education in Sri Lanka for Systems Change and Futures Literacy View Project PACSA: Phosphorus and Climate Smart Agriculture View Project. Available online: https://www.fao.org/fileadmin/user_upload/faoweb/ffc/docs/Climate%2520Resilient%2520CRFS%2520in-depth%2520assessment_IWMI.pdf (accessed on 20 July 2023).

Feintrenie, Laurène, Frank Enjalric, and Jean Ollivier. 2015. Coconut and Cocoa-Based Agroforestry Systems in Vanuatu: A Diversification Strategy in Tune with the Farmers' Life Cycle. In *Economics and Ecology of Diversification*. Dordrecht: Springer, pp. 283–95. [CrossRef]

Franzel, Steven, Sammy Carsan, Ben Lukuyu, Judith Sinja, and Charles Wambugu. 2014. Fodder Trees for Improving Livestock Productivity and Smallholder Livelihoods in Africa. *Current Opinion in Environmental Sustainability* 6: 98–103. [CrossRef]

Gautam, Madan K., Donald J. Mead, Peter W. Clinton, and Scott X. Chang. 2003. Biomass and Morphology of Pinus Radiata Coarse Root Components in a Sub-Humid Temperate Silvopastoral System. *Forest Ecology and Management* 177: 387–97. [CrossRef]

Hernández-Morcillo, Mónica, Paul Burgess, Jaconette Mirck, Anastasia Pantera, and Tobias Plieninger. 2018. Scanning Agroforestry-Based Solutions for Climate Change Mitigation and Adaptation in Europe. *Environmental Science and Policy* 80: 44–52. [CrossRef]

Hilimire, Kathleen. 2011. Integrated Crop/Livestock Agriculture in the United States: A Review. *Journal of Sustainable Agriculture* 35: 376–93. [CrossRef]

Ibrahim, M. N. M., and T. N. Jayatileka. 2000. Livestock Production under Coconut Plantations in Sri Lanka: Cattle and Buffalo Production Systems. *Asian-Australasian Journal of Animal Sciences* 13: 60–67. [CrossRef]

Jose, Shibu. 2009. Agroforestry for Ecosystem Services and Environmental Benefits: An Overview. *Agroforestry Systems* 76: 1–10. [CrossRef]

Kaplinsky, Raphael, and Mike Morris. 2018. Standards, Regulation and Sustainable Development in a Global Value Chain Driven World. *International Journal of Technological Learning, Innovation and Development* 10: 322. [CrossRef]

Kreitzman, Maayan, Mollie Chapman, Keefe O. Keeley, and Kai M. A. Chan. 2022. Local Knowledge and Relational Values of Midwestern Woody Perennial Polyculture Farmers Can Inform Tree-crop Policies. *People and Nature* 4: 180–200. [CrossRef]

Kumar, B. Mohan, and T. K. Kunhamu. 2022. Nature-Based Solutions in Agriculture: A Review of the Coconut (*Cocos nucifera* L.)-Based Farming Systems in Kerala, 'the Land of Coconut Trees. *Nature-Based Solutions* 2: 100012. [CrossRef]

Leakey, Roger R. B. 2017. Definition of Agroforestry Revisited. *In Multifunctional Agriculture* 8: 5–6. [CrossRef]

Lehmann, Johannes, Deborah A. Bossio, Ingrid Kögel-Knabner, and Matthias C. Rillig. 2020. The Concept and Future Prospects of Soil Health. *Nature Reviews Earth and Environment* 1: 544–53. [CrossRef] [PubMed]

McElwee, Pamela, Katherine Calvin, Donovan Campbell, Francesco Cherubini, Giacomo Grassi, Vladimir Korotkov, Anh Le Hoang, Shuaib Lwasa, Johnson Nkem, Ephraim Nkonya, and et al. 2020. The impact of interventions in the global land and agri-food sectors on Nature's Contributions to People and the UN Sustainable Development Goals. *Global Change Biology* 26: 4691–721. [CrossRef] [PubMed]

Mizik, Tamás. 2021. Climate-Smart Agriculture on Small-Scale Farms: A Systematic Literature Review. *Agronomy* 11: 1096. [CrossRef]

Nair, Kodoth Prabhakaran. 2021. Technological Advancements in Coconut, Arecanut and Cocoa Research: A Century of Service to the Global Farming Community by the Central Plantation Crops Research Institute, Kasaragod, Kerala State, India. In *Tree Crops*. Cham: Springer International Publishing, pp. 377–536. [CrossRef]

Nair, P. K. Ramachandran, B. Mohan Kumar, and S. Naresh Kumar. 2018. Climate Change, Carbon Sequestration, and Coconut-Based Ecosystems. *The Coconut Palm (Cocos nucifera L.)-Research and Development Perspectives*, 779–799. [CrossRef]

Nicholls, Clara I., and Miguel A. Altieri. 2013. Plant Biodiversity Enhances Bees and Other Insect Pollinators in Agroecosystems. A Review. *Agronomy for Sustainable Development* 33: 257–274. [CrossRef]

Nuwarapaksha, Tharindu D., Shashi S. Udumann, Nuwandhya S. Dissanayaka, Lakmini Dissanayake, and Anjana J. Atapattu. 2022. Coconut Based Multiple Cropping Systems: An Analytical Review in Sri Lankan Coconut Cultivations. *Circular Agricultural Systems* 2: 1–7. [CrossRef]

Nuwarapaksha, Tharindu D., Udaya N. Rajapaksha, Jayampathi Ekanayake, Senal A. Weerasooriya, and Anjana J. Atapattu. 2023a. Exploring the Economic Viability of Integrating *Jamnapari* Goat into Underutilized Pastures under Coconut Cultivations in Coconut Research Institute, Sri Lanka. *Biology and Life Sciences Forum* 27: 27. [CrossRef]

Nuwarapaksha, Tharindu D., Nuwandhya S. Dissanayaka, Shashi S. Udumann, and Anjana J. Atapattu. 2023b. Gliricidia as a Beneficial Crop in Resource-Limiting Agroforestry Systems in Sri Lanka. *Indian Journal of Agroforestry* 25: 12–18.

Paramesh, Venkatesh, Natesan Ravisankar, UmaKant Behera, Vadivel Arunachalam, Parveen Kumar, Racharla Solomon Rajkumar, Shiva Dhar Misra, Revappa Mohan Kumar, A. K. Prusty, D. Jacob, and et al. 2022. Integrated Farming System Approaches to Achieve Food and Nutritional Security for Enhancing Profitability, Employment, and Climate Resilience in India. *Food and Energy Security* 11: 321. [CrossRef]

Perera, Buthgamu M. A. O., and Marshall C. N. Jayasuriya. 2008. The Dairy Industry in Sri Lanka: Current Status and Future Directions for a Greater Role in National Development. *Journal of the National Science Foundation of Sri Lanka* 36: 115. [CrossRef]

Ramachandran Nair, P. K., Vimala D. Nair, B. Mohan Kumar, and Julia M. Showalter. 2010. Carbon Sequestration in Agroforestry Systems. *Advances in Agronomy* 108: 237–307. [CrossRef]

Raveendra, Thilina, Sarath P. Nissanka, Deepakrishna Somasundaram, Anjana J. Atapattu, and Sylvanus Mensah. 2021. Coconut-Gliricidia Mixed Cropping Systems Improve Soil Nutrients in Dry and Wet Regions of Sri Lanka. *Agroforestry Systems* 95: 307–19. [CrossRef]

Raveendra, Thilina, Anjana J. Atapattu, Sumith H. S. Senarathne, C. S. Ranasinghe, and Lasantha Weerasinghe. 2017. Evaluation of the Carbon Sequestration Potential of Intercropping Systems under Coconut in Sri Lanka. *International Journal of Geology, Earth and Environmental Sciences* 7: 1–7.

Rout, Pramod Kumar, and Basanta Kumara Behera. 2021. Cattle and Buffaloes Farming. In *Sustainability in Ruminant Livestock*. Singapore: Springer, pp. 77–115. [CrossRef]

Ruba, Umama Begum, and Mohammad Samiul Ahsan Talucder. 2023. Potentiality of homestead agroforestry for achieving sustainable development goals: Bangladesh perspectives. *Heliyon* 9: e14541. [CrossRef]

Samarajeewa, A. D., J. B. Schiere, M. N. M. Ibrahim, and Theo Viets. 2003. Livestock Farming in Coconut Plantations in Sri Lanka: Constraints and Opportunities. *Biological Agriculture and Horticulture* 21: 293–308. [CrossRef]

Sekaran, Udayakumar, Liming Lai, David A. N. Ussiri, Sandeep Kumar, and Sharon Clay. 2021. Role of Integrated Crop-Livestock Systems in Improving Agriculture Production and Addressing Food Security—A Review. *Journal of Agriculture and Food Research* 5: 100190. [CrossRef]

Seresinhe, Thakshala, and I. W. A. S. Sujani. 2016. Productivity Enhancement in a Cattle-Coconut Integrated System-Implications for Environmental Sustainability. Paper presented at the International Seminar on Livestock Production and Veterinary Technology, Jalan Raya Pajajaran Bogor, Indonesia, August 10–12; pp. 127–31. [CrossRef]

Sharma, Rashmitha, Usha Mina, and B. Mohan Kumar. 2022. Homegarden agroforestry systems in achievement of Sustainable Development Goals. A review. *Agronomy for Sustainable Developmen* 42: 44. [CrossRef]

Soler, Rosina, Pablo Luis Peri, Héctor Bahamonde, Verónica Gargaglione, Sebastián Ormaechea, Alejandro Huertas Herrera, Laura Sánchez Jardón, Cristian Lorenzo, and Guillermo Martínez Pastur. 2018. Assessing Knowledge Production for Agrosilvopastoral Systems in South America. *Rangeland Ecology and Management* 71: 637–45. [CrossRef]

Surendra, K. C., Devin Takara, Andrew G. Hashimoto, and Samir Kumar Khanal. 2014. Biogas as a Sustainable Energy Source for Developing Countries: Opportunities and Challenges. *Renewable and Sustainable Energy Reviews* 31: 846–59. [CrossRef]

Thornton, Philip K., and Mario Herrero. 2014. Climate Change Adaptation in Mixed Crop–Livestock Systems in Developing Countries. *Global Food Security* 3: 99–107. [CrossRef]

Utomo, Budi, Adi A. Prawoto, Sébastien Bonnet, Athikom Bangviwat, and Shabbir H. Gheewala. 2016. Environmental Performance of Cocoa Production from Monoculture and Agroforestry Systems in Indonesia. *Journal of Cleaner Production* 134: 583–91. [CrossRef]

Weerasinghe, Piyatilak. 2019. Livestock Feeds and Feeding Practices in Sri Lanka In Vitro and In Vivo Screening of Newly Introduced Forages for Sustainable Intensification of Dairy Production in the Context of Climate Change View Project Livestock Feeds and Feeding Practices in Sri Lanka. pp. 186–206. Available online: https://www.researchgate.net/publication/337971752 (accessed on 22 July 2023).

Wulandari, Sri Yulina. 2021. Bioentrepreneurship Approach as a Pillar to Accelerate the Integrated Farming System Implementation. *IOP Conference Series: Earth and Environmental Science* 733: 012107. [CrossRef]

Zoysa, Mangala De, and Makoto Inoue. 2014. Climate Change Impacts, Agroforestry Adaptation and Policy Environment in Sri Lanka. *Open Journal of Forestry* 4: 439–56. [CrossRef]

Approaches to Limiting Food Loss and Food Waste

Ioana Mihaela Balan, Teodor Ioan Trasca, Ioan Brad, Nastasia Belc, Camelia Tulcan, Bogdan Petru Radoi and Alexandru Erne Rinovetz

1. Introduction—Why and How Are the Quantities of Food Intended for Human Consumption Reduced?

Dynamics of food security have evolved over time, being influenced by demographic, economic, political, and technological changes. In general, the pre-industrial period was characterized by limited food security, with low agricultural production and high dependence on weather conditions (FAO et al. 2020, 2021; Pingali 2006).

With the development of agriculture and technology during the Industrial Revolution, food production increased significantly and food security began to improve. However, since the 20th century, climate change, population growth, and increased urbanization have brought new challenges to food security. In the 1960s and 1970s, the Green Revolution introduced new agricultural technologies, such as the use of chemical fertilizers and pesticides, to increase food production in developing countries. This revolution has brought about a significant improvement in food security in many countries. However, in recent decades, food security has become increasingly threatened by climate change, rising food prices, food loss and food waste, poverty, and conflict. In addition, the COVID-19 pandemic has brought new challenges to food security by disrupting food supply chains and increasing food prices. Currently, there is an increased need to address these challenges and develop sustainable solutions to ensure food security for all people (FAO et al. 2020; Godfray et al. 2010; Hoddinott and Haddad 1995; Parfitt et al. 2010; Pingali 2006).

According to the United Nations Environment Program (UNEP) and Food and Agriculture Organization (FAO) 2021 report, in 2020, approximately 768 million people worldwide were severely food insecure, representing approximately 9.9% of the global population. People experiencing food insecurity are found in all regions of the world, but most are in poor or developing countries. According to FAO data, the highest rates of food insecurity are in sub-Saharan Africa, where about a fifth of the population faces the problem, followed by South Asia (about 14%) and Caribbean and Central America (about 7.6%) (FAO et al. 2020, 2021; Boyd et al. 2019; UN WFP 2023).

The World Food Programme (WFP) showed in 2020 that, of all food produced for human consumption globally, a third is lost or wasted. In 2020, this represented 1.3 billion tons of food loss and food waste, which cost over USD 1 trillion. Moreover, all of this food which was produced but never eaten would have been more than enough to feed two billion people. This is more than twice the number of undernourished and malnourished people on the entire planet. Every year, the population of developed countries wastes almost as much food as the entire net annual food production of sub-Saharan Africa. However, in low-income and developing countries, 40% of food loss occurs at the post-harvest/slaughter/catch and processing levels. From an environmental impact perspective, looking cumulatively at what global food loss and food waste mean in relation to environmental impact, if the total of these quantities were a country, this country would be the third largest producer of carbon dioxide in the world, after the USA and China (UN WFP 2023).

The world is facing many interconnected crises. The triple planetary crisis of climate change, nature and biodiversity loss, and pollution and waste is accelerating (United Nations Climate Change 2022). Meanwhile, the war in Ukraine and other protracted conflicts are raising the prices of staple grains and threatening food security in many countries. These crises undermine efforts to achieve the Sustainable Development Goals (Ben Hassen and El Bilali 2022).

A shocking amount of food loss and food waste contributes to these crises, year after year. Even if estimates of their level differ from one source to another, due to the difficulties in collecting data from underdeveloped countries and some developing countries, categorically these quantities are enormous and worrying, especially in terms of their dynamics from one year to another. UNEP and FAO estimated in 2022 that 14% of the total food produced for human consumption is lost, while 17% is wasted. In the same report, UNEP and FAO noted in 2022 that food loss and food waste is sufficient to feed more than one billion people in a world where currently 828 million people are hungry and three billion cannot afford a healthy diet (FAO et al. 2022).

The link between food loss and food waste in terms of food security is an issue widely debated and analyzed by all international organizations dealing with food security concerns. The peculiarities of food systems in different countries and geographical areas determine the level of food loss and food waste, in the sense that they differ primarily according to the economic conditions of the country, geographical area, and a number of other factors which refer to the respective states (political, social conditions, presence of natural disasters, epidemics, etc.).

On the other hand, problems of food waste and food loss often present together, which can generate confusion and implicitly incorrect results of analyses and studies.

This is why it is necessary that presentation and analysis of current situations and intervention measures to reduce food loss and waste be treated separately, because these two categories, both of which can reduce the amount of food available for consumption, have different causes and implicitly different solutions. Thus, approaches to food loss and food waste must be different.

The situation can be clarified by a simple analysis of the presence of food loss and food waste along the food chain, in different food systems, depending on the particularities of states. If food loss prevails in underdeveloped countries, it manifests itself in the first stages of the food chain. If food waste prevails in developed countries, it manifests itself in the last stages of the food chain (UN WFP 2021) (Figure 1).

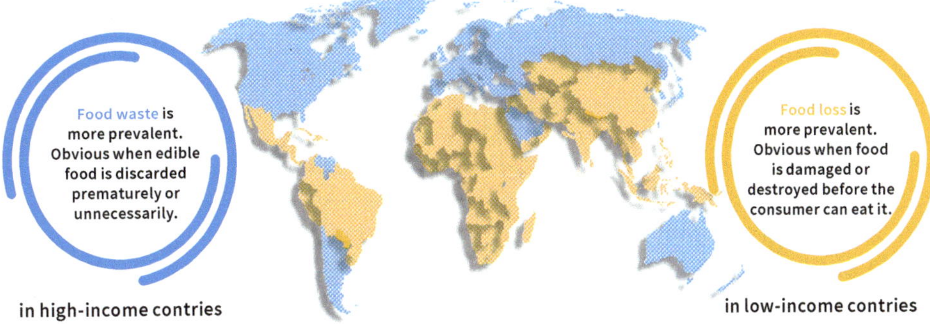

Figure 1. Prevalence of food loss and food waste by geographic area. Source: Authors' interpretation of UN WFP (2021).

The aims of the study on food loss and waste presented in this chapter have important meanings. First of all, it has the role of highlighting and quantifying the extent of the problem. By identifying the amount of food that is lost or wasted and by understanding the causes and mechanisms involved, a clear global picture of impact on the environment, economy, and society can be achieved (Morone et al. 2019b).

Another aim is to identify and assess factors that contribute to food waste, such as consumption habits, infrastructure, distribution systems, and specific policies. This understanding is crucial to develop effective strategies and solutions to reduce food waste. This study can provide recommendations and guidelines for authorities, companies, organizations, and consumers to help them adopt sustainable practices and promote effective food management.

The significance of this study consists of multiple aspects. First, reducing food waste and loss has a direct impact on global food security. By using food resources more efficiently, we can ensure that food is available for all and reduce pressure on the environment and agricultural systems.

Additionally, the study of food waste has important economic implications. Food loss and waste is a direct waste of resources, capital, and labor. By reducing them, we can increase economic efficiency and generate a more circular and sustainable economy (Morone et al. 2018).

Finally, the study of food loss and waste also has an ethical dimension. With approximately one-third of the world's food production being lost or wasted while millions of people suffer from hunger and malnutrition, it is essential to address this issue and for the population to take responsibility for using food resources wisely, in a fairer and more sustainable way.

Thus, the ultimate goal of this chapter is to promote a more sustainable, equitable, and efficient food system that ensures adequate nutrition for all, protects the environment, and contributes to sustainable economic and social development.

2. Materials and Methods

In this book chapter, we used an interdisciplinary approach, addressing the topic of food loss and food waste from multiple perspectives, including economic, scientific, and cultural–religious. This interdisciplinary approach and analysis of specialized literature in the field of food loss and waste represents useful academic research tools for examining complex topics and providing a comprehensive perspective of food security.

The materials accessed in this book chapter consist of relevant and credible studies, articles, and data collections. The research methods were the analysis of external secondary and tertiary data, collected by accessing scientific databases such as Web of Science, Publons, ResearchGate, and Google Scholar, as well as data provided by international organizations such as the United Nations, Food and Agriculture Organization, World Food Program, International Fund for Agricultural Development, United Nations International Children's Emergency Fund, World Bank, Our World in Data and other relevant organizations, the main websites of religious denominations, and religious texts in electronic or printed format. Additionally, the collected data and studies of the authors were accessed, reinterpreted and corroborated, as were the results of our own published research, which were published in scientific journals. All these data were analyzed, compared, and discussed in order to synthesize as faithfully as possible the approaches to limiting food loss and food waste.

3. Food Loss—Concept, Typology, Current Situation Worldwide

Food loss occurs when food inevitably becomes unfit for human consumption before people can eat it. The prevalence is higher in lower-income countries due to specific factors, when food is damaged or destroyed unintentionally by pests or mold (UN WFP 2021).

In low-income countries, food waste is low, and food loss predominates. In these countries, the population has a low income, limited access to food and other resources, and cannot afford to waste food. However, the lack of adequate infrastructure for food production and quality generates food loss in the first stages of the food chain, which subsequently generates situations of food insecurity. Although the situation is as serious as that of food waste, with percentages being comparable, the difference is the nature of the causes. In the case of food waste, this is generally a voluntary expression of the consumer; in the case of food loss, there is an involuntary character, which cannot be controlled, more and more often, due to lack of material resources, which reduces the ability to protect food production.

In terms of relation to the food chain, food loss is manifested in the first stages, namely in the agricultural production phase, storage phase, and then in food processing phase, as well as in the distribution phase (Figure 2).

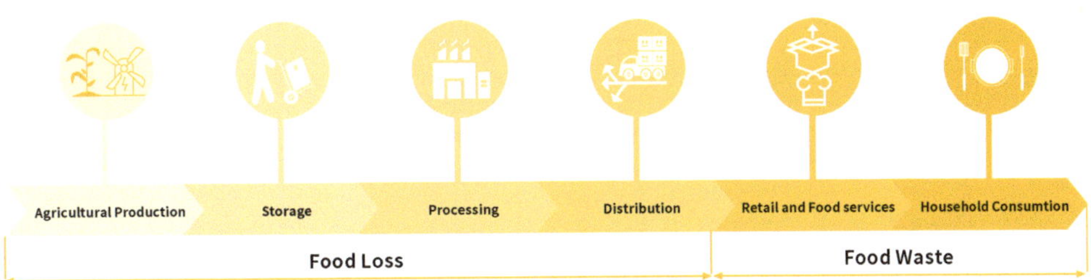

Figure 2. Prevalence of food loss and food waste in the food chain. Source: Authors' interpretation of FAO (2011a, 2011b).

Unlike food loss, which has involuntary human causes, food waste manifests itself in the last stages of the food chain as a voluntary attitude, namely in the retail trade and food service phase as well as in the food consumption phase at the household level. By exception, the retail phase may exhibit both forms of food reduction, namely food loss and food waste. This is due, in terms of food loss, to improper food display and marketing conditions. On the other hand, food waste can manifest itself in this retail phase of the food chain as a result of the aesthetic demands of consumers. Thus, various foods such as fruits or vegetables remain in grocery stores or agro-food markets due to aesthetic defects that have nothing to do with their nutritional value.

In low-income countries, and sometimes not only in these countries, food can be lost as a result of the harvesting process taking place prematurely. Farmers with low incomes sometimes harvest crops too early, and the motivation is given by food deficiency or acute need for money, from the second half of the agricultural season. In this way, food loses its nutritional and economic value and can often be wasted if it is not suitable for consumption (FAO 2011a, 2011b).

Poor storage facilities and a lack of infrastructure cause post-harvest/slaughter/catch food loss in developing countries. Many farms in low-income countries are not connected to electricity grids and have no other source of electricity, so storage conditions (temperature, humidity) are unsuitable for storing and preserving food. In addition, in many of these countries, the hot and humid climate competes to create a storage environment that is difficult to ensure in the absence of electricity sources. Fresh foods such as vegetables, fruits, meat, milk, eggs, and fish directly from the farm or after catch can be damaged in hot and/or excessively humid climates due to a lack of transport, storage, cooling infrastructure, and agri-food markets (FAO 2011a, 2011b).

According to UN's World Food Program USA (UN WFP 2021), causes of food loss are diverse, but the most important are:

- Food lost before harvest due to drought or storms;
- Crops infested with insects, resulting from improper storage;
- Food spoiled during transport due to lack of refrigeration.

As for the post-harvest/slaughter/catch loss, they vary according to their category (Table 1).

Table 1. Share of food loss in post-harvest/slaughter/catch processes.

Type of Food	% of Loss
Cereals and pulses	8.6%
Fruits and vegetables	21.6%
Meat and animal products	11.9%
Other	10.1%
Roots and tubers	25.3%

Source: FAO (2019).

Use of cold technologies in the development of agricultural supply chains for meat, dairy products, fish, and vegetable products began in the early 1950s with the growth of the mechanical refrigeration industry. However, cold chains are still very limited in most developing countries (Meat.Milk 2022). There are many technical, logistical, and investment challenges as well as economic opportunities related to the use of cold chain. The main segments of an integrated cold chain include (Meat.Milk 2022)

- Packaging and cooling of fresh food products;
- Food processing (i.e., freezing certain processed foods);
- Cold storage (short- or long-term storage of refrigerated or frozen food);
- Distribution (cold transport and temporary storage at controlled temperature);
- Merchandising (cold or freezer storage and displays in wholesale markets, retail markets and food service operations).

Currently, a considerable part of this loss is caused by improper cold chain processes and management. However, worldwide, about a third of fresh fruit and vegetables are thrown away because their quality has fallen below an acceptable limit, which is totally unacceptable. Much of this loss is related to incorrect handling during supply chain processes. However, conceptually, the notion of "Expiry Date" is outdated; it only refers to number of days a food product is of "acceptable quality" and safe to eat, and these requirements depend on the conditions of optimum temperature and transport (Jedermann et al. 2014). A better concept, however, would be that of "First in—first out", which was introduced in the late 1980s. The idea behind the concept is to apply stock rotation so that each product's remaining shelf life best matches the remaining transport duration options, reducing product waste during transport, and ensuring product consistency in the store (Jedermann et al. 2014).

Food quality variations and remaining shelf life are automatically calculated from accumulated data on environmental conditions such as temperature variations and shelf life, which are then used by warehouse management software to match shelf life variation, stock rotation storage, routing and special handling. Unfortunately, due to a lack of automatic data capture and shelf life calculation systems, this much more efficient concept has found very few practical applications to date (Jedermann et al. 2014).

4. Food Waste—Concept, Typology, Current Situation Worldwide

In industrialized countries, food is lost when production exceeds demand. To ensure the delivery of agreed quantities while anticipating bad weather or unpredictable pest attacks, farmers sometimes make safe production plans and end up producing larger quantities than needed, even when conditions are "average". If they produced more than needed, some surplus crops are sold to processors or as animal feed. However, this is often not financially viable given the lower prices in these sectors compared to retailers (FAO 2011a, 2011b).

Globally, about 14% of food produced is lost between harvest and retail, while an estimated 17% of total global food production is wasted: 11% in households, 5% in food services, and 2% in retail (United Nations 2022). However, the forms of food waste can be diverse if also analyzed from the perspective of the consumer. Thus, metabolic food waste is a real form of food waste, characterized by an overconsumption of food. Metabolic food waste has the same characteristics as any other form of food waste, i.e., it wastes food without needing to, thus reducing the amount of food that could be effectively consumed (Balan et al. 2022a).

On the other hand, unsafe food is not fit for human consumption and is therefore wasted. Failure to meet minimum food safety standards can lead to food loss and, in extreme cases, impact a country's food security status. A number of factors can lead to food insecurity, such as naturally occurring toxins in food, contaminated water, unsafe use of pesticides, and veterinary drug residues. Poor and unsanitary handling and storage conditions and a lack of adequate temperature control can also cause unsafe food (Toma et al. 2020).

Whether real or just perceived, food safety is one of the most important reasons for food waste, throughout the agri-food supply chain. On-farm food safety risks such as mycotoxin contamination of feed, overuse of antimicrobials in animal disease control, and the incursion of zoonotic diseases can lead to food unfit for human consumption and thus waste. Given the importance of food safety as one of the most important attributes of all food, proper risk management along the supply chain can help reduce food loss and waste. Therefore, there is a need to improve the coordination between food waste and food safety policies. This new approach to coordination must necessarily involve balancing scientific evidence and the precautionary principle. At the same time, it is necessary to review current regulations on food safety to identify those areas that generate waste that can be avoided. There is also a need for hazard monitoring combined with waste monitoring along the entire agri-food chain, followed by the provision of tailored and realistic information on the links between food safety and waste. Last but not least, investments in high-performance technologies are needed to accurately assess the degree of edibility of food. At the same time, increased attention is required on food labeling and packaging policies and practices. In their absence, it cannot be guaranteed that food labeling and packaging do not generate unintended and unnecessary impacts on food safety and, by implication, food waste, which are not justified by scientific evidence (Toma et al. 2020).

5. Distribution of Food Loss and Food Waste along the Food Chain and Measures to Reduce Them

The prevalence of food loss and waste varies according to the stage of the food chain. As shown above, in low-income countries, food loss in the early stages of the food chain is greater than food waste in the later stages of the food chain. This is diametrically opposite in high-income countries, where there is less food loss in the early stages of the food chain and more food waste in the later stages of the food chain. These variations directly depend on food systems and the level of technology, as well as the demands and requirements of the consumer, compared to the global level.

5.1. *Analyses of the Levels of Food Loss and Food Waste*

The level of food loss and waste is analyzed in order to achieve SDGs, especially SDG 12: Ensure sustainable consumption and production patterns, Indicator 12.3.1: (a) Food loss index and (b) food waste index. Globally, the situation of these two indicators shows variations depending on a number of factors which are present in different geographical areas (Our World in Data 2009) (Figure 3).

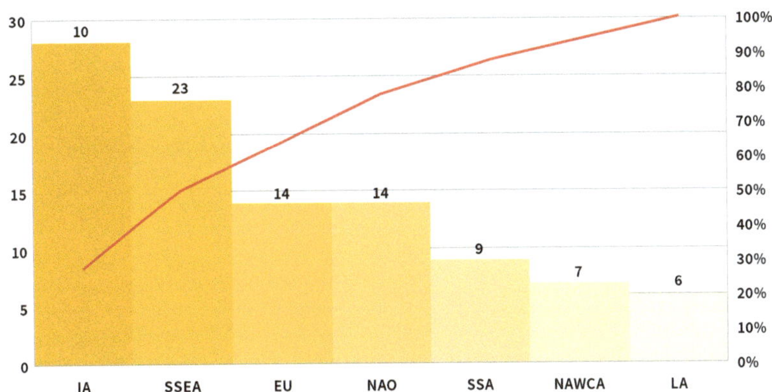

Figure 3. Share of global food loss and waste by region. Source: Authors' interpretation of Our World in Data (2009). Legend: IA—Industrialized Asia; SSEA—South and Southeast Asia; EU—Europe; NAO—North America and Oceania; SSA—Sub-Saharan Africa; NAWCA—North Africa, West, and Central Asia; LA—Latin America.

It is noteworthy that over 50% of global food loss and waste is recorded in Asia. This fact is due to the large population, but also to specific conditions along the food chain. Europe and North America, together, present the same values as industrialized Asia. This fact shows that, although they are countries with a high level of industrialization, concerns and the effects of measures to reduce food loss and food waste are low (Lin et al. 2014).

The level of food waste recorded in 2019, depending on the segments of the food chain that involve retail and individual consumption, varies significantly depending on country, although they are interpretable depending on data collection methods.

The global reported level of food loss and waste in retail varies between 3.12 and 78.82 kg/year/capita, with the highest values in Malaysia, and lowest in New Zealand (Our World in Data 2019) (Figure 4).

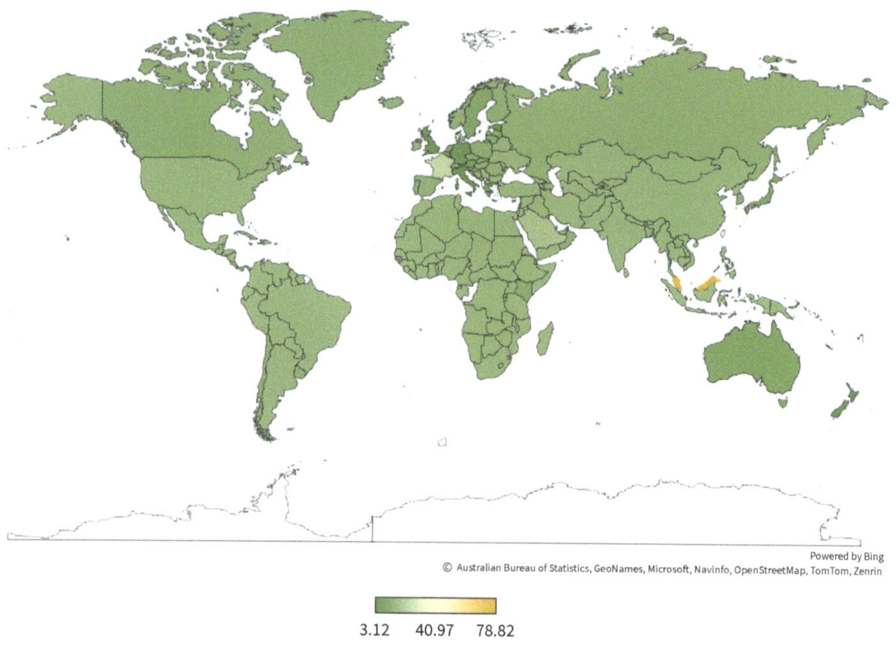

Figure 4. The level of retail food waste. Source: Authors' interpretation of Our World in Data (2019).

Additionally, the global reported level of household food waste varied between 33.38 and 164.36 kg/year/capita, with the highest values in Rwanda, and the lowest in Russia (Our World in Data 2019) (Figure 5).

The global level of out-of-home food waste varies between 3.34 and 89.56 kg/year/capita, with the highest values in Malaysia and the lowest in Bangladesh (Our World in Data) (Figure 6).

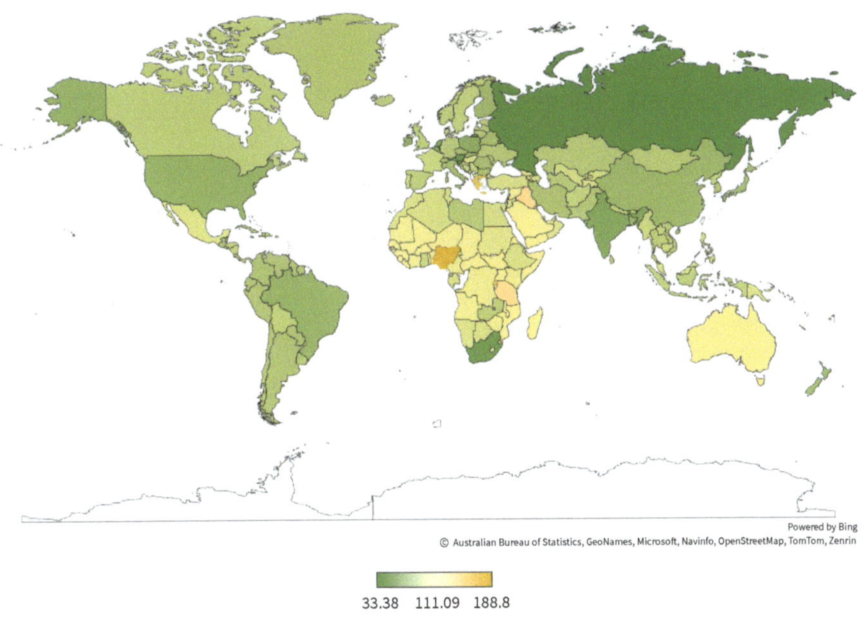

Figure 5. The level of households food waste. Source: Authors' interpretation of Our World in Data (2019).

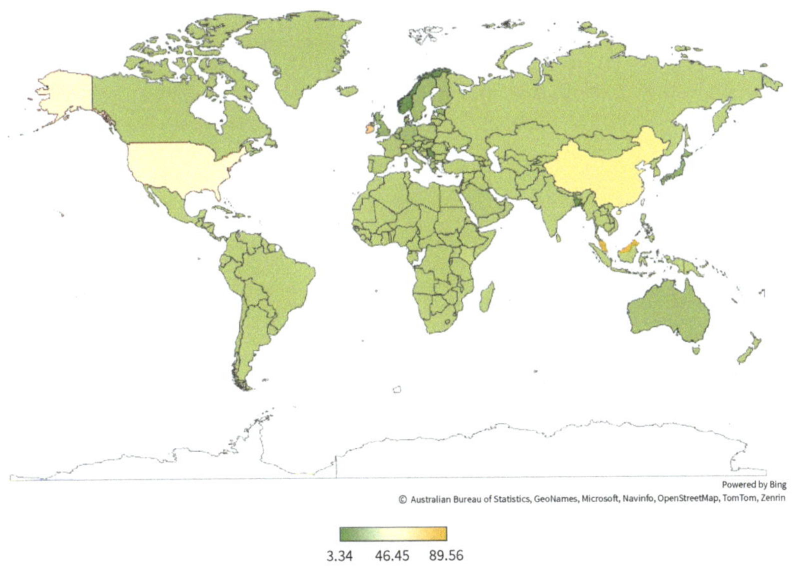

Figure 6. The level of out-of-home food waste. Source: Authors' interpretation of Our World in Data (2019).

The paradoxical situation of food consumption and waste is not only characteristic to some specific countries, but is a problem faced by many countries all over the world. Even in countries with high levels of food consumption, food waste remains a major challenge. This phenomenon can be attributed to common factors such as consumption habits, poor infrastructure, and insufficient education in effective food management (Falcone and Imbert 2017; Garcia-Herrero et al. 2018; Lin et al. 2014; Otles and Kartal 2018) (Box 1).

Box 1. The Paradox of Food Waste in Romania.

At the national level, Romania, located in Southeast Europe, can be considered a representative example for the study of food loss and food waste, as well as for food security, due to the importance of the agricultural sector in economy, food diversity and authenticity, consumption behavior, infrastructure, and government policies. The study of these aspects can contribute to the development of effective strategies and policies to reduce waste and improve food security not only in Romania, but also globally. (Dumitru et al. 2021)

Romania is a country with an average level according to several criteria. From an economic point of view, Romania has registered a significant growth in recent decades, but it is still in the process of developing to reach the level of developed countries. Its GDP, although growing, is still below the European average. However, Romania has considerable economic potential, with rich natural resources and a skilled workforce, which can contribute to sustainable economic growth in the future.

The population of Romania is significant, being around 19 million inhabitants. Compared to other European countries, Romania has a relatively low population density, which can provide opportunities for development in certain areas. With a territorial area of approximately 238,397 square kilometers, Romania ranks as one of the largest countries in Central and Eastern Europe. This territorial expansion offers considerable potential for economic development and for geographical and cultural diversity. (World Countries n.d.)

An important aspect which has to be considered when analyzing Romania as an average country is food waste and loss. Romania faces challenges in efficient management of food resources and in reducing waste. Despite being an agricultural country with significant food production, an important part of this food is wasted before it reaches the final consumers. Efforts to reduce food loss and promote the more efficient management of food resources represent important challenges for the Romanian government and society as a whole. (Dumitru et al. 2021)

Romania has a strong agricultural tradition and rich natural resources, which makes it dependent on the agricultural sector. A good part of the economy is based on agriculture and food production. Thus, it is important to understand and analyze how production, distribution, and consumption processes are carried out in order to identify areas where food loss and waste occur. On the other hand, Romania enjoys a rich culinary diversity and a variety of traditional and local food products. This provides opportunities to study how these products are preserved and promoted, but also risks losing traditional knowledge and unique food resources. (IRCEM 2021)

Related to food consumption and behavior, the study of food waste and loss can also focus on the consumption habits of the population. Romania is experiencing changes in lifestyle and food preferences, and understanding how these influence the amount of wasted food can help to identify effective strategies to reduce them.

In Romania, infrastructure and food distribution systems play an important role in food waste. The study of Romania can highlight aspects related to storage, transport, and logistics, as well as how they can be improved to reduce food loss and waste. (IRCEM 2021; FEBA and CHEP 2016)

Government policies and initiatives of the Romanian Government involve measures to address the problem of food waste and loss. Policies and initiatives are implemented to promote food security, reduce waste, and improve food resource management systems. A Romanian case study can assess the effectiveness of these measures and provide examples and lessons for other countries. (Gheorghescu and Balan 2019a) In this context, a study from 2019 highlights that approximately 33% of the Romanian population spends 30–40% of their monthly income on food, while 29% of citizens allocate 20–30% of their monthly income for this purpose. Despite this fact, over 50% of food waste and loss occurs at the households level (Figure 7). (IRCEM 2021; Gheorghescu and Balan 2019b)

Romania faces a paradoxical situation in terms of food consumption and waste. Although the population spends a significant proportion of their monthly income on food relative to their standard of living, food waste remains a major problem. This can be attributed to several factors, including eating habits, poor infrastructure, and lack of education on effective food management. (Gheorghescu and Balan 2019a) One of the main reasons for the high food waste in Romania is related to consumption habits and food preferences. Sometimes, people buy more food than they can eat and it ends up being wasted. There is also a tendency to throw away food on its expiry date due to food safety concerns, even though it could still be eaten. Analyzing from another perspective, enormous amounts of food are wasted in Romania, and the population in a state of moderate and severe food insecurity has increased in recent years (Figure 8). (World Food Programme–Hunger Map 2022)

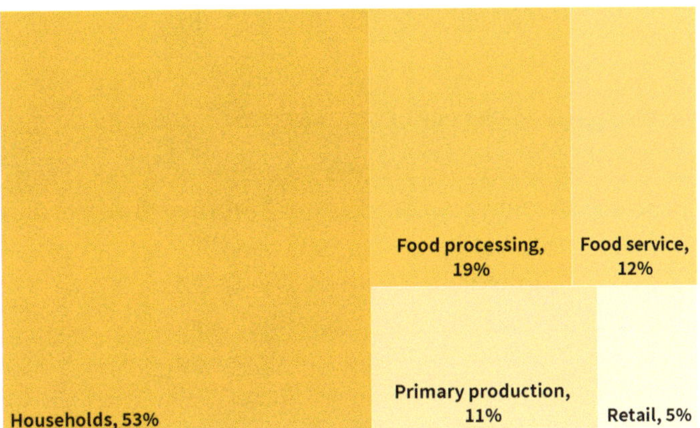

Figure 7. The distribution of food loss and food waste in the Romanian food chain. Source: Authors' interpretations of Gheorghescu and Balan (2019b).

In addition, poor infrastructure and an inefficient distribution system also contribute to food waste. Sometimes food spoils before reaching consumers due to deficiencies in supply chain, storage, or transportation. This leads to a loss of significant amounts of food, which could be avoided by better developed infrastructure and more efficient distribution systems.

To combat food waste in Romania, an integrated and coordinated approach is needed, which includes education and awareness among the population, improving infrastructure and distribution systems, as well as promoting efficient food management practices throughout the supply chain. Thus, Romania can reduce food loss and contribute to a more sustainable use of food resources, bringing benefits to the environment, economy, and well-being of the population. (Chereji et al. 2023; Gheorghescu and Balan 2019a)

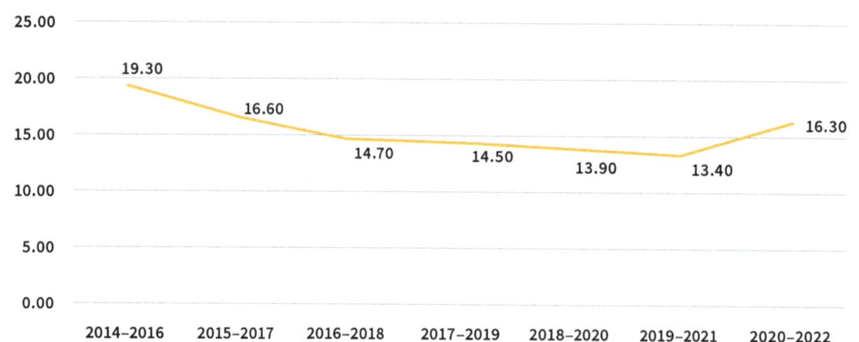

Figure 8. Prevalence of moderate or severe food insecurity in the total population in Romania (percent) (3-year average) Source: Authors' interpretation of World Food Programme–Hunger Map (2022).

5.2. Distribution of Food Loss and Food Waste along the Food Chain

Production: Global loss at this stage is estimated at 20–40%, with considerable variation depending on the type of agricultural product. The main causes include pest and disease infestation, adverse climatic conditions, poor crop management, and deficiencies in infrastructure and logistics (FAO 2011a, 2011b).

Storage: Food loss at this stage can reach 10–20% globally. These are mainly due to improper storage conditions, such as inadequate temperatures or humidity, pest infestation, and improper handling of the product (FAO 2011a, 2011b).

Processing: Food loss at this stage varies with food product, but has been estimated at 5–20% globally. This can be caused by product mishandling, damage during transport, interruption of the cold chain, and improper storage (FAO 2011a, 2011b).

Distribution: Food loss at this stage is estimated at 10–15% globally, with significant variations by region and product type. These losses can be caused by poor transport and handling conditions, problems with infrastructure and logistics, and problems with documentation and import/export formalities (FAO 2011a, 2011b).

Retail and food services: This stage can be responsible for food loss of up to 5–10% globally, but also for food waste due to excess packaging, unsold products or products that expire on the shelf. Uneven marketing of products can also be a problem, leading to oversupply or undersupply. In food service, food waste can be caused by too-large portions or a failure to recover leftover food (FAO 2019). Food loss can occur at this stage due to product degradation during handling and transport, but also the withdrawal from sale of products that are no longer fresh or past their expiry date. There is also food waste due to unsold products or those that are thrown away for other reasons.

Household: According to the FAO, around 40% of global food waste occurs at the household level (FAO 2019). This includes food that is thrown away or that is damaged and thrown away, as well as food that is eaten as leftovers or leftover food that is thrown away.

This finding is confirmed by several studies and reports from around the world. For example, a study by the UK's Waste and Resources Action Program (WRAP) found that around 70% of food thrown away in the UK comes from households (WRAP 2021). Another study in Ontario, Canada, found that households are responsible for about 47% of all food waste in that region (Parfitt et al. 2010).

In general, household food waste can have many causes, such as overbuying food, improper food storage, past expiration dates, or preparing too much food. For this reason, taking steps to reduce food waste at household level can be a key point in waste reduction efforts. It is important to note that these figures are only estimates and food waste can be difficult to measure accurately in some regions.

The enormous amounts of food that are lost and wasted have prompted international organizations to try to reduce them for several years. Thus, numerous studies have been developed through which different key measures, as well as complementary measures, have been identified in order to reduce them. As we have shown before, the problem of humanity today is not the lack of food resources, but precisely the problem of food that, although it was intended for human consumption, never ends up being consumed (Falcone and Imbert 2017; Gheorghescu and Balan 2019a; Morone et al. 2019a, 2019b; Otles and Kartal 2018).

5.3. Measures to Reduce Food Loss and Food Waste along the Food Chain

Related to the food chain, these key measures and complementary measures can be synthesized according to the stage in which they take place. (FAO 2019; Parfitt et al. 2010; O'Donnell et al. 2015; Balan et al. 2022b; Meat.Milk 2022) (Figure 9).

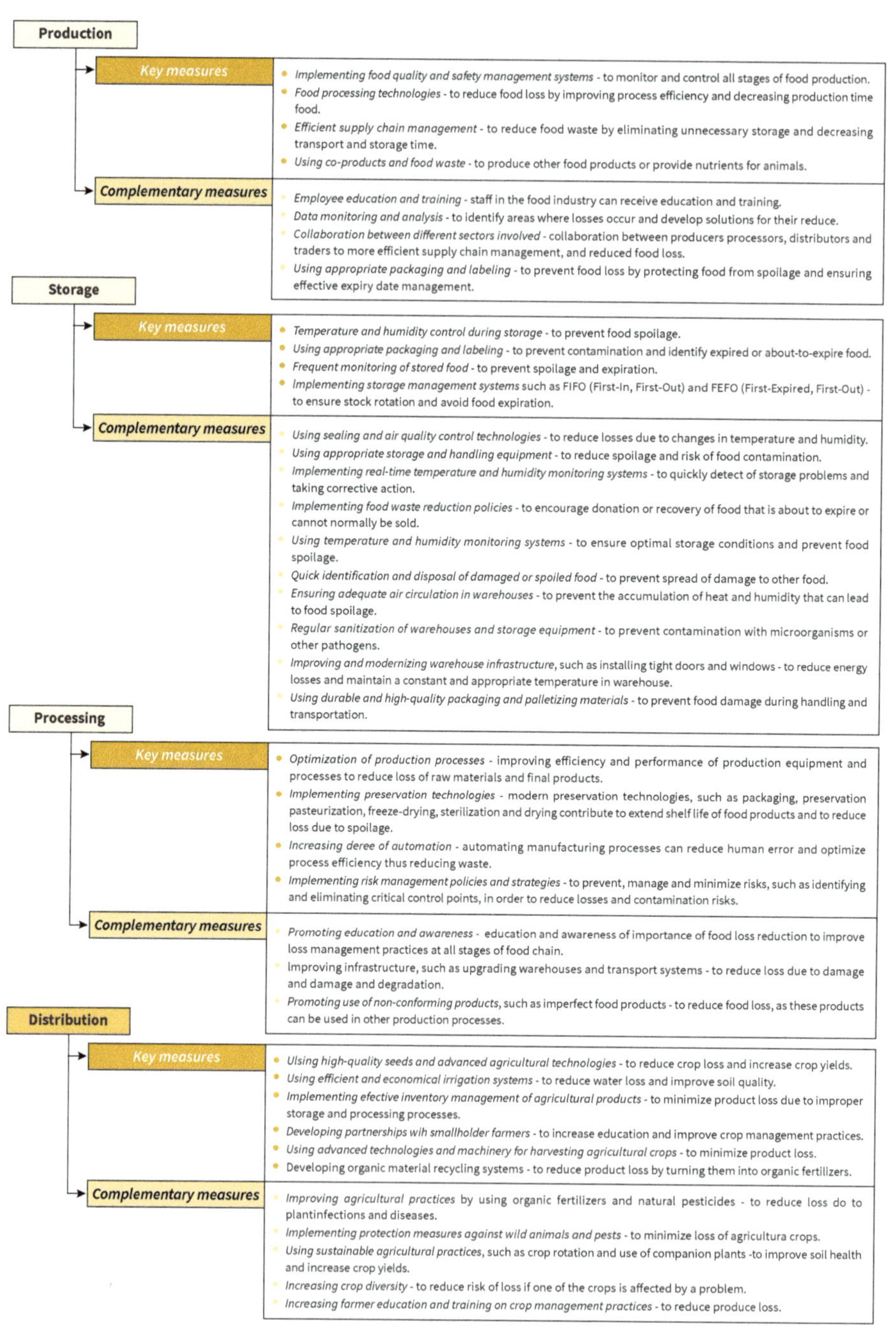

Figure 9. *Cont.*

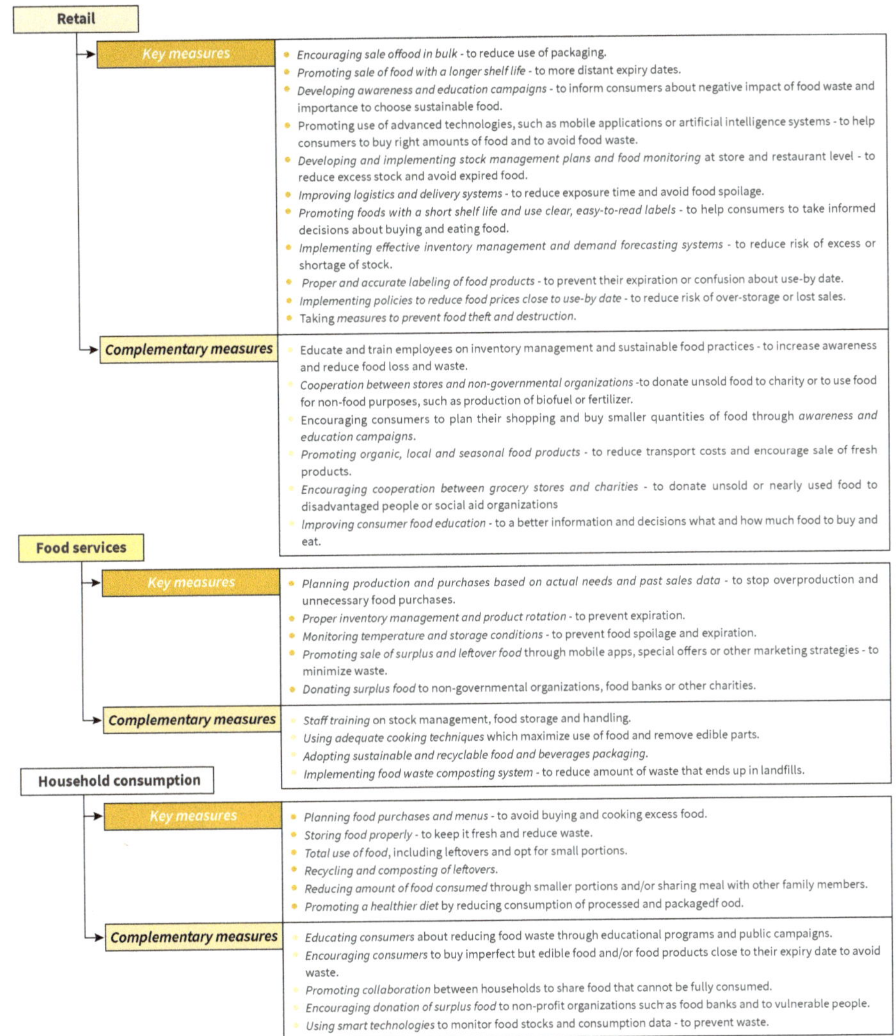

Figure 9. Key measures and complementary measures to reduce food loss and food waste along the food chain. Source: (FAO 2019; Parfitt et al. 2010; O'Donnell et al. 2015; Balan et al. 2022b; Meat.Milk 2022).

6. The Impact of Food Loss and Food Waste on the Environment

The further along the chain that food loss occurs, the more carbon-intensive the waste. For example, fruit and vegetables spoiled at the harvest stage will have a lower carbon footprint than jams or canned vegetables wasted at retail store, because harvesting, transport, and processing accumulate additional greenhouse gases along the supply chain (FAO 2011a, 2011b).

The environmental impact of food loss and waste can be quite significant and can have long-term negative effects on our planet. Here are some concrete examples: (Lipinski et al. 2013; FAO 2019; Kummu et al. 2012; Parfitt et al. 2010)

Greenhouse effect: When food is thrown away and decomposes, it releases greenhouse gases such as carbon dioxide and methane into the atmosphere. These gases contribute to global warming and climate change.

Energy and water consumption: Producing food requires a significant amount of energy and water, and when food is thrown away, that energy and water is wasted. In addition, the food production process can deplete natural resources such as soil and water.

Land development and deforestation: To produce food, land must be cultivated and maintained, which can lead to deforestation and overdevelopment of agricultural land. This can lead to the loss of natural habitats for animals and plants, as well as soil and water pollution.

On global average, the per capita climate footprint of food waste in high-income countries is more than double that of low-income countries, due to wasteful food distribution and consumption patterns in high-income countries (Table 2) (FAO et al. 2020).

Table 2. Per capita food wastage CO_2 footprint on climate.

Regions	kg CO_2/Capita/Year
North America and Oceania	860
Industrialized Asia	810
Europe	680
Latin America	540
North Africa, Western Asia, Central Asia, South and Southeast Asia	350
Sub-Saharan Africa	210

Source: FAO et al. (2020).

As previously shown, if food waste were analyzed as a country, this country would be the third largest emitter of CO_2 in the world, after China and the USA (FAO 2013).

Globally, food loss and food waste generate 4.4 GtCO_2 equivalent annually, or about 8% of total anthropogenic greenhouse gas (GHG) emissions. This means that the contribution of wasted food emissions to global warming is almost equivalent to global emissions from road transport, i.e., 87% of them. (FAO 2013) At the same time, the volume of water used for them is equivalent to that of Lake Geneva, and the agricultural area used is approximately 30% of the total agricultural area (Figure 10).

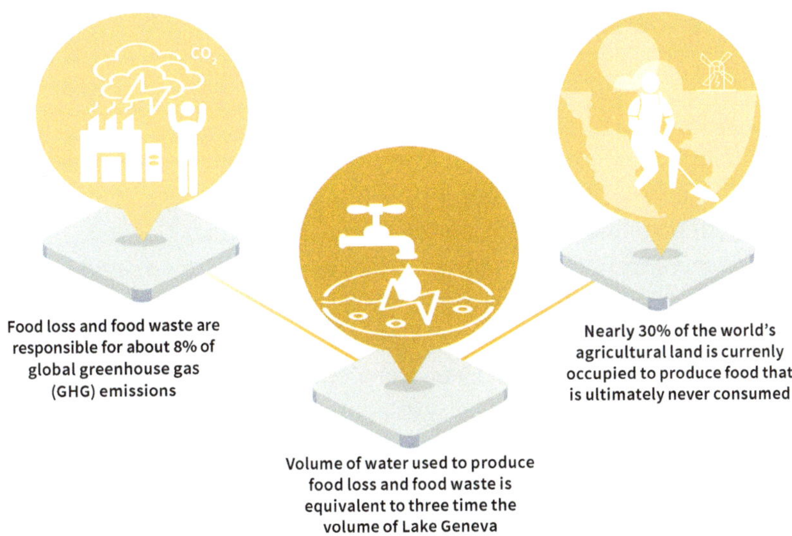

Figure 10. Environmental impact of food loss and food waste. Source: Authors' interpretation of Geneva Environment Network (2022).

7. Ethics Regarding Food Loss and Food Waste

Ethically, food loss and waste also have significant implications. From the perspective of inequitable food distribution, food waste exacerbates inequality and poverty around the world. Food loss and food waste reduce the total amount of food available, which can lead to higher prices and increase the difficulty of accessing food for those who are already socio-economically vulnerable.

The environmental impact is felt by the entire population, both those who waste food and those who do not waste food. Food waste contributes to environmental pollution through greenhouse gas emissions and the production of food waste that ends up in landfills, affecting both those who waste food and those who ethically manage food resources and avoid waste.

At the same time, the limitation of resources also affects the entire population. Food loss and waste is an inefficient use of the planet's limited resources. Food production involves the use of many natural and financial resources such as water, land, energy, and workforce. Wasting and losing food constitutes an inefficient exploitation of these resources, which could be used for other important purposes (Gustavsson et al. 2011; Buzby et al. 2014; Kantor et al. 1997).

In terms of business ethics, companies that contribute to food waste can be perceived as unethical. These companies not only make financial losses, but also contribute to negative environmental impact and social problems. Such companies could be seen as prioritizing their profits over ethical values such as social and environmental responsibility (Gustavsson et al. 2011).

Overall, food loss and food waste are complex ethical issues with negative environmental and societal implications. Solving these problems requires a strong commitment from governments, companies, and individual consumers to improve the efficiency and sustainability of the food system (Gustavsson et al. 2011; Buzby et al. 2014; Kantor et al. 1997; De Schutter 2014; Lipinski et al. 2013).

On the other hand, at the religious level, since ancient times, humanity has been concerned with reducing food waste. However, with the rise of technology and, implicitly, modern man, humanity's responsibility towards food and towards all resources necessary to obtain it (water, land, workforce, energy) has obviously decreased. However, throughout history, the situation has not always been the same as it is today. It is important to note that, in general, all religions and spiritual traditions emphasize the importance of respect for food and nourishment and encourage moderation and generosity.

In Christianity, biblical passages in the Old and New Testaments that refer to waste and respect for food and foodstuffs are numerous. Perhaps the most representative is "When they were all full, Jesus said to his disciples: "*Gather up the pieces that are left over. Let nothing be wasted.*" (John 6:12. The Holy Bible; https://www.biblegateway.com/passage/?search=John%206&version=NIV, accessed on 2 March 2023)

In Quran it is stated precisely and representatively: "*Eat and drink, but do not waste, for He does not love the wasteful.*" (Quran 7:31; https://quran.com/7/31?translations=20,44,17,85,18,95,48,39,26,101,41,19,22,38,31,27,33, accessed on 2 March 2023)

In one of Judaism's most important texts, the Talmud, there are several references to the importance of avoiding food waste. Perhaps most eloquent in this context are "*It is forbidden to throw away food or destroy it on purpose*" (Talmud,

Shabbat 128a; https://www.sefaria.org/Shabbat?tab=contents, accessed on 2 March 2023) and *"No one should throw away food, but should feed it to domestic and wild animals."* (Talmud, Baba Metzia 62b).

Some references in canonical Buddhist texts regarding reduction in food waste are: *"Food should never be wasted, but should be treated as if it were nectar"* (Vinaya Pitaka, Part I, Sutta 3; https://www.britannica.com/topic/Vinaya-Pitaka, accessed on 20 September 2023), *"We should recognize the efforts and work done by people to produce the food that we eat and do not waste it"* (Majjhima Nikaya, Sutta 55; https://www.britannica.com/topic/Majjhima-Nikaya, accessed on 20 September 2023), and *"We should be careful not to waste food and not we treat it with contempt, because every morsel of food is a valuable treasure"* (Sutta Nipata, Sutta 103; https://www.britannica.com/topic/Suttanipata, accessed on 20 September 2023).

In Hindu religious texts, there are also references to food waste and the importance of respecting food. Here are some examples and information about where to find them in the respective texts: *"If you waste food, you also waste the life within you"* (Mahabharata, Book 13, Anusasana Parva, Section XLVIII; https://www.britannica.com/topic/Mahabharata, accessed on 20 September 2023), *"Do not waste food, because it is the source of your life and wealth"* (Yajurveda, Taittiriya Upanishad, Shiksha Valli, Anuvaka 11; https://www.britannica.com/topic/Yajurveda, accessed on 20 September 2023), and *"We should cherish food and not waste it, because it is the source of life and vitality"* (Bhagavadgita, chp. 17, Verse 10; https://www.britannica.com/topic/Bhagavadgita, accessed on 20 September 2023).

These examples emphasize the importance of valuing and respecting food, and that food is source of life and wealth. In the Hindu religion, food is considered a form of divine gift and should therefore be treated with respect and gratitude.

In the Sikh religion, food is considered sacred and should be respected as such. There is the concept of "langar", which is a kind of community kitchen where food is provided for free and everyone is welcome to eat together, regardless of caste or social status (Nesbitt 2016).

In the Taoist tradition, great importance is placed on food and nutrition, believing that food is essential to the health of body and mind. The consumption of natural and whole foods is promoted and food waste is avoided (Maoshing and McNease 2012).

In Confucian philosophy, food and eating are seen as an act of respect for oneself and others. Respect and appreciation of food is promoted, and food waste is seen as an undesirable behavior (Eno 2015).

These are just a few examples of religions and spiritual traditions that place great importance on food and avoid food waste.

8. Conclusions

Food loss and waste is a global problem with a significant impact on society in general, with economic, social, and ecological consequences. The economic effects of food loss and waste are significant, as they increase the cost of food production, transport, and storage. Furthermore, food loss and food waste reduce the profitability of food businesses and ultimately affect food prices for end consumers.

The social impact of food loss and waste is also important, as it can lead to hunger and malnutrition, especially in low- and middle-income countries. Food loss and waste also contribute to economic and social inequality by reducing access to quality and affordable food.

In terms of ecological impact, food loss and food waste are a major environmental problem. They generate greenhouse gas emissions and unnecessary use of natural resources such as water, land, and energy, thus contributing to climate change and environmental degradation. In this context, reducing food loss and waste is a global priority and requires immediate action at all levels. These actions may include improving food production, transport, and storage systems, educating consumers about the importance of reducing food loss and food waste, and promoting technological innovations to improve efficiency and sustainability of the food chain.

Initiatives and programs aimed at reducing food loss and food waste are diverse and at different levels. There are many initiatives globally that aim to reduce food loss and food waste at the production level by improving production and distribution systems and promote more sustainable practices, in the context of the UN's Sustainable Development Goal 2—Zero Hunger (SDG 2—Zero Hunger). An example of such an initiative is the UNEP, which launched a program called "Think.Eat.Save" to promote a reduction in food waste globally (United Nations 2022). There are also numerous consumer campaigns and initiatives that try to promote a more responsible attitude towards food and reduce food waste. These include education and awareness programs, as well as campaigns to reduce food waste at the level of households and public institutions. An example of such an initiative is "Save Food, Fight Waste", an awareness campaign launched by the Swiss Government to promote reduction in food waste (Government of Switzerland 2022). In addition, there are also non-governmental organizations and volunteer groups working to reduce food waste by collecting and redistributing uneaten food. An example of such an organization is Feeding America (https://www.feedingamerica.org/about-us, accessed on 28 December 2022), a US charity that collects uneaten food and redistributes it to those who need it.

Food loss and waste is a complex problem and efforts are being made at all levels to reduce it. However, the initiatives are insufficient. It is important that actions focus on finding viable, efficient, and effective solutions to reduce food loss and waste and promote more efficient and sustainable use of food resources, in the context to assure social, economic, and environmental necessary conditions (Figure 11).

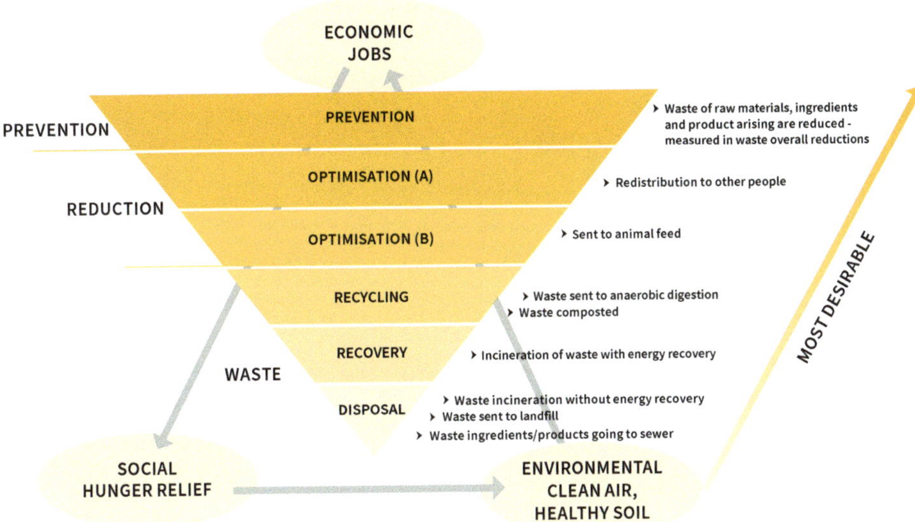

Figure 11. Hierarchy of food recovery. Source: Authors' interpretation of CEC (2023).

The FAO's 2011 assessment of global food loss and waste estimated that, each year, one-third of all food produced worldwide for human consumption never reaches the consumer's table. This is not only a missed opportunity for the economy and food security, but also a waste of all the natural resources used to grow, process, pack, transport, and market food. Through an extensive literature search, a 2011 assessment of food waste volumes gathered weighted ratios of food loss and waste for different regions of the world, and examined different commodity groups and different stages of the supply chain. These ratios have been applied to regional food mass flows in FAO's Food Balance Sheets 2007. Food waste occurs at all stages of the food supply chain for a variety of reasons that dependent largely on the local conditions in each country. Globally, a pattern is clearly visible; in high-income regions, volumes of wasted food are higher in the processing, distribution, and consumption stages, while in low-income countries, food loss occurs in the production and post-harvest/slaughter/catch stages.

Future lines of research on food loss and waste should focus on several key aspects. First, it is important to continue to develop and improve methods for quantifying levels of food loss and waste (Morone et al. 2019a). Current methods can often be limited by accuracy and difficulties in collecting relevant data. Therefore, researchers should explore new approaches and technologies that allow a more accurate and efficient estimation of the amount of food loss and waste at different stages of the food chain.

Another important aspect is the analysis of the causes and factors that contribute to food waste. A deeper understanding of consumption habits, food preferences, production practices, and structural factors that can influence food waste is needed. This could involve researching consumer behavior, food policy, distribution systems, and infrastructure to identify hotspots and effective solutions to reduce waste.

It is also important to assess the impact of food loss and waste on the environment, economy, and society. Research should focus on assessing food waste's carbon footprint, greenhouse gas emissions, water footprint, estimating associated economic and social costs, and analyzing impacts on food security and human health.

As we move forward in food waste research, we should also promote collaboration between different disciplines and sectors (Otles and Kartal 2018). Interdisciplinary research can bring innovative insights and approaches, such as integrating digital technologies and real-time data analytics to monitor and reduce food waste.

As a necessity that imposes itself, future research directions should focus on developing more accurate and efficient methods to quantify food waste, understanding the causes and factors involved, and assessing the environmental and economic impacts. By addressing these issues, we can make significant progress in reducing food waste and promoting a more sustainable and efficient food system.

As a wake-up call, humanity needs to take note that if food loss and food waste are not reduced worldwide, several negative things will happen:

- Increasing poverty and hunger: Food loss and food waste reduce the amount of food available to eat. This can increase food prices and make them unaffordable for those living in poverty, increasing the risk of hunger and malnutrition.

- Rising food prices: Food loss and food waste have a significant impact on food prices. If no action is taken to reduce this loss, it will continue to increase, leading to higher food prices.
- Impact on environment: Food loss and food waste have a significant impact on the environment. Food production involves the use of resources such as water, energy, soil, and others. If these resources are used inefficiently, more waste and greenhouse gas emissions will result, contributing to climate change.
- Social inequities: Food loss and food waste are often linked to social inequalities. While some regions and populations are affected by famine, others are constantly throwing away food. While some countries face problems of obesity and overweight, others face malnutrition and famine. Food waste compounds these problems, as food thrown away could be used to feed people suffering from malnutrition.
- Reducing food loss and food waste could help redistribute food to those in need.
- Health problems: Food loss and food waste can lead to health problems among the population through unbalanced or poor quality diets.
- Economic decline: Food loss and food waste represent a loss of economic resources, and a waste of money and resources for agricultural producers and other companies involved in food production.

In the context of SDG 2—Zero Hunger, the development of efficient and effective programs to reduce levels of food loss and waste can ensure more food for more people, provided they pursue equity and equality among all people in the world. At the same time, it will have a positive impact on the environment, productivity will increase, economic growth will be generated and societies will become more sustainable.

Author Contributions: Conceptualization, I.M.B. and T.I.T.; Methodology, I.B. and N.B.; Software, B.P.R. and A.E.R.; Validation, I.M.B. and T.I.T.; Formal Analysis, C.T., B.P.R., and A.E.R.; Investigation, I.M.B., T.I.T., and C.T.; Resources, B.P.R. and A.E.R.; Data Curation, C.T., B.P.R., and A.E.R.; Writing—Original Draft Preparation, I.M.B.; Writing—Review and Editing, T.I.T. and C.T.; Visualization, I.M.B., T.I.T., and C.T.; Supervision, I.B. and N.B.; Project administration, I.M.B.; Funding Acquisition, C.T. and N.B. All authors have read and agreed to the published version of the manuscript.

Funding: This research was performed in the frame of the project Development and consolidation of the METROFOOD-RI National Network, grant offered by the Romanian Minister of Research, Innovation, and Digitalization, as Intermediate Body for the Competitiveness Operational Program 2014–2020, call POC/78/1/2/, project number SMIS2014 + 136213, acronym METROFOOD-RO.

Acknowledgments: This research was performed in University of Life Sciences "King Mihai I of Romania"—Research Institute for Biosecurity and Bioengineering.

Conflicts of Interest: The authors declare no conflict of interest.

References

Balan, Ioana Mihaela, Emanuela Diana Gherman, Ioan Brad, Remus Gherman, Adina Horablaga, and Teodor Ioan Trasca. 2022a. Metabolic Food Waste as Food Insecurity Factor—Causes and Preventions. *Foods* 11: 2179. [CrossRef] [PubMed]

Balan, Ioana Mihaela, Emanuela Diana Gherman, Remus Gherman, Ioan Brad, Raul Pascalau, Gabriela Popescu, and Teodor Ioan Trasca. 2022b. Sustainable Nutrition for Increased Food Security Related to Romanian Consumers' Behavior. *Nutrients* 14: 4892. [CrossRef] [PubMed]

Ben Hassen, Tarek, and Hamid El Bilali. 2022. Impacts of the Russia-Ukraine War on Global Food Security: Towards More Sustainable and Resilient Food Systems? *Foods* 11: 2301. [CrossRef] [PubMed]

Boyd, A. Swinburn, Vivica I. Kraak, Steven Allender, Vincent J. Atkins, Phillip I. Baker, Jessica R. Bogard, Hannah Brinsden, Alejandro Calvillo, Olivier De Schutter, Raji Devarajan, and et al. 2019. The Global Syndemic of Obesity, Undernutrition, and Climate Change: The Lancet Commission report. *The Lancet Commissions* 393: 791–846. [CrossRef]

Buzby, Jean C., Hodan F. Wells, and Jeffrey Hyman. 2014. The Estimated Amount, Value, and Calories of Postharvest Food Losses at the Retail and Consumer Levels in the United States. In *Economic Information Bulletin*; Washington, DC: United States Department of Agriculture (USDA), p. 121. Available online: https://www.ers.usda.gov/webdocs/publications/43833/43680_eib121.pdf (accessed on 20 March 2023).

Commission for Environmental Cooperation (CEC). 2023. Seven Steps for Measuring Food Loss and Waste Within Your Business, City, STATE, or Country. Available online: http://www.cec.org/flwm/checklist-item/why-measure-food-loss-and-waste/ (accessed on 8 March 2023).

Chereji, Aurelia-Ioana, Irina-Adriana Chiurciu, Anca Popa, Ioan Chereji, and Adina-Magdalena Iorga. 2023. "Consumer Behaviour Regarding Food Waste in Romania, Rural versus Urban". *Agronomy* 13: 571. [CrossRef]

De Schutter, Oliver. 2014. Report of the Special Rapporteur on the Right to Food. United Nations General Assembly, Human Rights Council. Available online: https://digitallibrary.un.org/record/766914?ln=en (accessed on 20 March 2023).

Dumitru, Oana M., Corneliu S. Iorga, and Gabriel Mustatea. 2021. Food Waste along the Food Chain in Romania: An Impact Analysis. *Foods* 10: 2280. [CrossRef]

Eno, Robert. 2015. The Analects of Confucius. An Online Teaching Translation. (Version 2.21). Penguin Classics. ISBN: 978-0140443486. Available online: https://chinatxt.sitehost.iu.edu/Analects_of_Confucius_(Eno-2015).pdf (accessed on 10 March 2023).

Falcone, Pasquale Marcello, and Enrica Imbert. 2017. Bringing a Sharing Economy Approach into the Food Sector: The Potential of Food Sharing for Reducing Food Waste. In *Food Waste Reduction and Valorisation*. Edited by P. Morone, F. Papendiek and V. Tartiu. Cham: Springer. [CrossRef]

FAO. 2011a. Food Wastage Footprint & Climate Change. Available online: https://www.fao.org/3/bb144e/bb144e.pdf (accessed on 12 March 2023).

FAO. 2011b. Global Food Losses and Food Waste—Extent, Causes and Prevention. Available online: https://www.fao.org/sustainable-food-value-chains/library/details/en/c/266053/ (accessed on 12 March 2023).

FAO. 2013. Food Wastage Footprint: Impacts on Natural Resources. Summary Report. Available online: https://www.fao.org/3/i3347e/i3347e.pdf (accessed on 12 March 2023).

FAO. 2019. *The State of Food and Agriculture 2019: Moving Forward on Food Loss and Waste Reduction*. Rome: FAO. Available online: https://www.fao.org/3/ca6030en/ca6030en.pdf (accessed on 12 March 2023).

FAO, IFAD, UNICEF, WFP, and WHO. 2020. *The State of Food Security and Nutrition in the World 2020: Transforming Food Systems for Affordable Healthy Diets*. Rome: FAO. [CrossRef]

FAO, IFAD, UNICEF, WFP, and WHO. 2021. *The State of Food Security and Nutrition in the World 2021: Transforming Food Systems for Food Security, Improved Nutrition and Affordable Healthy Diets for All*. Rome: FAO. [CrossRef]

FAO, IFAD, UNICEF, WFP, and WHO. 2022. *The State of Food Security and Nutrition in the World 2022: Repurposing Food and Agricultural Policies to Make Healthy Diets More Affordable*. Rome: FAO. [CrossRef]

European Federation of Food Banks (FEBA), and Commonwealth Handling Equipment Pool (CHEP). 2016. Case Study: Fight Food Waste in Romania. Available online: https://www.chep.com/si/en/case-study-fight-food-waste-romania (accessed on 20 March 2023).

Garcia-Herrero, Isabel, Daniel Hoehn, María Margallo, Jara Laso, Alba Bala, Laura Batlle-Bayer, Pere Fullana-i-Palmer, Ian Vázquez-Rowe, M.J. Gonzalez, M.J. Durá, and et al. 2018. On the estimation of potential food waste reduction to support sustainable production and consumption policies. *Food Policy* 80: 24–38. [CrossRef]

Geneva Environment Network. 2022. Food Loss and Waste and the Role of Geneva. Available online: https://www.genevaenvironmentnetwork.org/resources/updates/reducing-food-loss-and-waste-for-a-healthier-planet/ (accessed on 1 March 2023).

Gheorghescu, Ionut Cosmin, and Ioana Mihaela Balan. 2019a. Managing, minimizing and preventing food waste from Romania in the European context. *Lucrări Stiintifice Management Agricol* 21: 21–58. Available online: https://lsma.ro/index.php/lsma/article/view/1707/pdf (accessed on 1 March 2023).

Gheorghescu, Ionut Cosmin, and Ioana Mihaela Balan. 2019b. Food waste and food loss in Romania. *Lucrări Științifice Management Agricol* 20: 356. Available online: http://www.lsma.ro/index.php/lsma/article/view/1429 (accessed on 20 March 2023).

Godfray, H. Charles J., John R. Beddington, Ian R. Crute, Lawrence Haddad, David Lawrence, James F. Muir, Jules Pretty, Sherman Robinson, Sandy M. Thomas, and Camilla Toulmin. 2010. Food Security: The Challenge of Feeding 9 Billion People. *Science* 327: 812–18. [CrossRef] [PubMed]

Government of Switzerland. 2022. Save Food, Fight Waste. Available online: https://www.savefood.ch/ (accessed on 28 December 2022).

Gustavsson, Jenny, Christal Cederberg, Ulf Sonesson, Robert van Otterdijk, and Alexandre Meybeck. 2011. *Global Food Losses and Food Waste*. Rome: FAO. Available online: https://www.fao.org/3/i2697e/i2697e.pdf (accessed on 20 March 2023).

Hoddinott, John, and Lawrence Haddad. 1995. Does Female Income Share Influence Household Expenditures? Evidence from Côte d'Ivoire. *Oxford Bulletin of Economics and Statistics* 57: 77–96. Available online: https://onlinelibrary.wiley.com/doi/10.1111/j.1468-0084.1995.tb00028.x (accessed on 20 March 2023). [CrossRef]

Institute of Research in Circular Economy and Environment (IRCEM). 2021. Study of Food Waste in Romania. Romanian Circular Economy Stakeholder Platform. Available online: https://rocesp.ro/wp-content/uploads/2022/02/V4-final-Food-Waste-Report.pdf (accessed on 21 March 2023).

Jedermann, Reiner, Nicometo Mike, Uysal Ismail, and Lang Walter. 2014. Reducing food losses by intelligent food logistics. *Philosophical Transactions of the Royal Society A* 372. [CrossRef] [PubMed]

Kantor, Linda Scott, Kathrin Lipton, Alden Manchester, and Victor Oliveira. 1997. *Estimating and Addressing America's Food Losses*; Washington, DC: United States Department of Agriculture (USDA). Available online: https://endhunger.org/docs_hunger/USDA-Jan97a.pdf (accessed on 20 March 2023).

Kummu, Matti, Hans de Moel, Miina Porkka, Stefan Siebert, Varis Olli, and Philip J. Ward. 2012. Lost food, wasted resources: Global food supply chain losses and their impacts on freshwater, cropland, and fertiliser use. *Science of the Total Environment* 438: 477–89. [CrossRef] [PubMed]

Lin, Carol Sze Ki, Apostolis A. Koutinas, Katerina Stamatelatou, Egid B. Mubofu, Avtar S. Matharu, Nikolaos Kopsahelis, Lucie A. Pfaltzgraff, James H. Clark, Seraphim Papanikolaou, Tsz Him Kwan, and et al. 2014. Current and future trends in food waste valorization for the production of chemicals, materials and fuels: A global perspective. *Biofuels, Bioproducts and Biorefining* 8: 686–715. [CrossRef]

Lipinski, Brian, Craig Hanson, Richard Waite, Tim Searchinger, and James Lomax. 2013. Reducing Food Loss and Waste. Working paper, World Resources Institute, Washington, DC, USA. Available online: https://www.wri.org/publication/reducing-food-loss-and-waste (accessed on 20 September 2023).

Maoshing, Ni, and Cathy McNease. 2012. *The Tao of Nutrition*, 3rd ed. Boulder: Shambhala Publications. Available online: https://www.pdfdrive.com/the-tao-of-nutrition-e25991207.html (accessed on 10 March 2023).

Meat.Milk. 2022. Reducing Food Loss through Efficient Use of the Cold Chain. Available online: https://www.meat-milk.ro/diminuarea-pierderilor-de-alimente-prin-utilizarea-eficienta-a-lantului-rece/ (accessed on 5 January 2023).

Morone, Piergiuseppe, Pasquale Marcello Falcone, Enrica Imbert, and Andrea Morone. 2018. Does food sharing lead to food waste reduction? An experimental analysis to assess challenges and opportunities of a new consumption model. *Journal of Cleaner Production* 185: 749–60. [CrossRef]

Morone, Piergiuseppe, Falcone Pasquale Marcello, and Tartiu Valentina Elena. 2019a. Food waste valorisation: Assessing the effectiveness of collaborative research networks through the lenses of a COST action. *Journal of Cleaner Production* 238: 117868. [CrossRef]

Morone, Piergiuseppe, Falcone Pasquale Marcello, and Lopolito Antonio. 2019b. How to promote a new and sustainable food consumption model: A fuzzy cognitive map study. *Journal of Cleaner Production* 208: 563–74. [CrossRef]

Nesbitt, Eleanor. 2016. *Sikhism: A Very Short Introduction*. Oxford: Oxford University Press. Available online: https://www.google.ro/books/edition/Sikhism_A_Very_Short_Introduction/zD8SDAAAQBAJ?hl=en&gbpv=1&dq=Nesbitt,+Eleanor.+2016.+Sikhism:+A+Very+Short+Introduction.+Oxford+University+Press,+ISBN:+978-0198745570&printsec=frontcover (accessed on 2 March 2023).

O'Donnell, T., J. Deutsch, R. Pepino, B.J. Milliron, C. Yungmann, and S. Katz. 2015. Food Was Never Meant to Be Wasted. *Food Waste*. October 21. Available online: https://www.biocycle.net/food-was-never-meant-to-be-wasted/ (accessed on 2 March 2023).

Otles, Semih, and Canan Kartal. 2018. 11—Food Waste Valorization. In *Sustainable Food Systems from Agriculture to Industry*. Edited by Charis M. Galanakis. Cambridge: Academic Press, pp. 371–99. [CrossRef]

Our World in Data. 2009. Share of Global Food Loss and Waste by Region, 2009. Available online: https://ourworldindata.org/grapher/global-food-waste-by-region (accessed on 12 May 2023).

Our World in Data. 2019. Food Waste Per Capita, 2019. Available online: https://ourworldindata.org/grapher/food-waste-per-capita (accessed on 12 May 2023).

Parfitt, Julian, Michael Barthel, and Sarah MacNaughton. 2010. Food Waste within Food Supply Chains: Quantification and Potential for Change to 2050. *Philosophical Transactions of the Royal Society B* 365: 3065–81. [CrossRef] [PubMed]

Pingali, Prabhu. 2006. Agricultural Growth and Economic Development: A View through the Globalization Lens. Paper presented at 26th International Conference of Agricultural Economists, Gold Coast, Australia, August 12–18; Available online: https://citeseerx.ist.psu.edu/document?repid=rep1&type=pdf&doi=795db66ba7c9cfbe293ef881857635e87706e0ba (accessed on 12 March 2023).

Toma, Luiza, Cesar Revoredo-Giha, Montserrat Costa-Font, and Bethan Thompson. 2020. Food Waste and Food Safety Linkagesalong the Supply Chain. *EuroChoices* 19: 24–29. [CrossRef]

UN WFP—World Food Program USA. 2021. Food Waste vs. Food Loss: Know the Difference and Help #StopTheWasteToday. Available online: https://www.wfpusa.org/articles/food-loss-vs-food-waste-primer/ (accessed on 2 March 2023).

UN WFP—World Food Program USA. 2023. Hunger Map. Available online: https://hungermap.wfp.org/?_ga=2.254518287.66435396.1679456688-1887669960.1679456688 (accessed on 20 March 2023).

United Nations. 2022. International Day of Awareness on Food Loss and Waste Reduction, 29 September. Available online: https://www.un.org/en/observances/end-food-waste-day (accessed on 8 March 2023).

United Nations Climate Change. 2022. What Is the Triple Planetary Crisis? Available online: https://unfccc.int/blog/what-is-the-triple-planetary-crisis (accessed on 20 March 2023).

World Countries—Romania. n.d. Available online: https://www.globecountries.com/country/romania.html (accessed on 1 June 2023).

World Food Programme–Hunger Map. 2022. Available online: https://www.fao.org/interactive/state-of-food-security-nutrition/2-1-1/en/ (accessed on 26 August 2023).

WRAP. 2021. Waste and Resources Action Programme (WRAP). Food Surplus and Waste in the UK—Key Facts. Available online: https://wrap.org.uk/resources/report/food-surplus-and-waste-uk-key-facts (accessed on 1 March 2023).

MDPI
St. Alban-Anlage 66
4052 Basel
Switzerland
Tel. +41 61 683 77 34
Fax +41 61 302 89 18
www.mdpi.com

MDPI Books Editorial Office
E-mail: books@mdpi.com
www.mdpi.com/books

www.ingramcontent.com/pod-product-compliance
Lightning Source LLC
LaVergne TN
LVHW071927160726
843515LV00010B/2542